SEED REGENERATION
IN CROSS-POLLINATED SPECIES

SEED REGENERATION IN CROSS-POLLINATED SPECIES

Proceedings of the C.E.C. / EUCARPIA Seminar
Nyborg / Denmark / 15-17 July 1981

Edited by

E.PORCEDDU
Istituto del Germoplasma, Bari, Italy

G.JENKINS
Agricultural Research Council, London, UK

A.A.BALKEMA / ROTTERDAM / 1982

Published for the Commission of the European Communities,
Directorate-General Information Market & Innovation, Luxembourg

EUR 7892 EN

ISBN 90 6191 244 X

Published by A.A.Balkema, P.O. Box 1675, 3000 BR Rotterdam, Netherlands
Distributed in USA & Canada by MBS, 99 Main Street, Salem, NH 03079

Text typed by Janssen Services, 33a High Street, Chislehurst, Kent BR7 5AE, UK
Printed in the Netherlands

Table of contents

Session 4. *Technical aspects: Field crops*

Final session

Preface

The Symposium reported in this publication was organised by the Committee for Genebank and Resistance Breeding of the Directorate General for Agriculture of the Commission of the European Communities, in collaboration with the Section 'Wild Species and Primitive Forms' of EUCARPIA.

The premise must be accepted that regeneration of genebank stocks with minimal loss of genetic integrity is an important aspect of the preservation of genetic resources in which there is a lack of modern studies and research. Samples of self-pollinated species can be easily regenerated or multiplied by harvesting seeds produced in nursery rows adjacent to other introductions of the same species. This method does not hold good for increasing seed lots of cross-pollinated species as most of the harvested seeds will be hybrids of unknown male parentage and will not be representative of the introduction.

Consequently, the purpose of this Symposium was to discuss methods to increase seed stocks of cross-pollinated species, their scientific implication (number of plants, isolation, method of inter-crossing, etc.) in different species with the aim of formulating an acceptable strategy.

The Symposium was organised in four sessions dealing respectively with (1) Theoretical Aspects, (2) Practical Aspects in regeneration of field crop plants, (3) Practical Aspects in regeneration of pasture plants, (4) Practical Aspects in regeneration of horticultural plants. Four invited papers and 15 reports were discussed in the meeting and are included in the Proceedings.

Opening

E. Porceddu *(Italy)* On behalf of the EC and EUCARPIA, I wish to convey our sincere appreciation to Professor Andersen for his warm welcome to us in this garden of Denmark and for his active interest in preparing this meeting.

I also wish to thank you all for accepting the invitation to attend this meeting, in spite of the many commitments which I know you have at your experimental stations at this time of the year. Your presence here in such great numbers is an indication of the importance of the subject and justifies our initial conception of this meeting some two or three years ago, first in the EC and later in EUCARPIA.

We are all aware that plant germplasm is one of man's most valuable natural resources. It provides genetic material for modification, improvement and breeding of high yielding varieties, and has also proved indispensable to agricultural development through the production of improved, well-adapted and disease resistant cultivars in many crops.

The task of collecting and conserving these natural genetic resources has been the responsibility of national and international genetic resource centres around the world. Most collections consist of a limited number of seeds initially and therefore require multiplication prior to storage and characterisation. Seed distribution to scientists and seed ageing requires regeneration of seed lots. Genetic integrity of this germplasm material must be preserved during this activity for which suitable methods have been devised in the case of self-pollinated crops such as barley, wheat and rice. The integrity of the sample in self-pollinated species can be preserved by collecting and multiplying single seeds as separate sub-samples of an original sample.

However, such methods are not appropriate in the case of

cross-pollinated species such as forages, maize, rye, beet, vegetables, etc. which represent a large number of crops spread all over the world, because most of the harvested seeds will be hybrids of unknown male parentage and may not be equivalent to the original sample.

In cross-pollinated species, each plant is different from all the other plants in the population and from those in the next generation. Population samples are identified by genotype and gene frequencies and their regeneration requires the unchanged preservation of these frequencies. We all know that the ideal solution would be to grow accessions in their natural habitat. However, that is almost impossible, either because the population has been removed from its original location or because the agricultural conditions have changed since the time when the accessions were collected, and thus the population has adapted to a new environment. We should adopt any device aimed at avoiding change in the genetic composition of the population.

The problem is not dissimilar to that encountered in the maintenance of breeders' seed and in the multiplication of this seed for commercialisation of a new cultivar. The methods of regeneration of cross-pollinated species and their implication in different species need detailed discussion and the formulation of a practical strategy for curators of collections. It is the purpose of this meeting to canvass your views on this subject which is of great importance to the Community. I hope that some solutions to the problems involved will be forthcoming during the course of this meeting.

With this aim, I now declare the Symposium open.

J.G. Hawkes *(UK)* I would like to welcome you also on behalf of EUCARPIA, Section: Wild Species and Primitive Forms, which is collaborating with the EC Genebank Group in the organisation of this meeting. EUCARPIA has been able to invite participants from outside the EC countries and I would like to extend a very warm welcome to those people as well as to the EC participants. On a subject like this I think it is important to have the widest possible participation throughout Europe in order that we may arrive at views and a consensus of opinion based on scientific research and arguments as to the best ways of regeneration of seeds in cross-pollinated species. As Professor Porceddu has said, the collections of material for breeding are now becoming more and more important within Europe, and indeed, throughout the whole world, in providing basic stocks for plant breeding development and improvement of varieties. These collections, normally known as seed banks or genebanks, are ones that are basic to the work of plant breeders. Every collection has its limitation in the number of years it can be stored before it finally becomes non-viable, or begins the process of losing its viability, and before the need for regeneration becomes very apparent. We must, at all costs, try to conserve the genetic integrity of the sample so as to prevent some change in the population structure every time it is regenerated. As Professor Porceddu pointed out, this is not too difficult in self-pollinated crops, and perhaps even easier in vegetatively propagated material. However, there are problems in cross-pollinated species, particularly of preserving the allelic frequency of populations, which means that we have to take certain measures for cross-pollinated crops that are not necessary for others. Both in EUCARPIA and in the EC Group, it has been felt that this is a matter which has not been fully discussed previously and we believe that it is very important to do this in order to provide guidelines for research workers in the field. We hope that by the end of this meeting we will have taken some reasonably clear decisions about the best ways of dealing with the various types of cross-pollinated crops. We have to consider numbers of plants for regeneration, not only to prevent inbreeding depression, but also to prevent genetic drift in the populations. For this reason we need to make sure that the numbers are adequate in our regeneration plots. Furthermore, we have to take particular care to prevent contamination of the material by alien pollen, pollen from other gene pools, from other collections, because in a genebank or in a series of collections, we may be trying to regenerate several samples at the same time and there must be some ways of isolating one sample from another so that its own pollen is used in the regeneration of seeds and pollen from other samples is prevented from contaminating that sample. So, the preservation of the gene pool is a very important topic which I hope we will discuss, particularly in relation to the preservation of the genetic integrity of each sample. The ideal ways may be very expensive and time consuming, depending on the different crops, the different nature of the material, and for this reason we have divided the sessions into groups of crops so that the different solutions for the various groups can be clearly discussed and understood.

Session 1
Theoretical aspects

Chairman: J.G.HAWKES

Theoretical problems of seed regeneration in cross-pollinated plants

Y.DEMARLY
Université de Paris-Sud, Orsay, France

INTRODUCTION

During recent decades many genebanks for autogamous plants or vegetatively propagated species have been established throughout the world. In a few cases these dealt with cross-pollinated plants but they were usually stores for seed samples or plant samples and have not solved the many problems of preservation of the genetic diversity of this kind of species over a long period of time.

This situation must be due to some characteristic features of allogamous species and particularly to their sensitivity in the genotype of each plant and of the population structure to its environmental conditions.

First, we shall analyse the organisation of these systems. Secondly, the maintenance systems of such populations in their original areas of production will be presented. Then we shall see what kind of methods could be used not only to prevent their genetic erosion but also to give access to their wealth.

CHARACTERISTIC ASPECTS OF BREEDING SYSTEMS IN PLANTS

At the population level

As our knowledge of evolutionary genetics progresses, we cannot consider the reproductive pattern of a species without integrating a series of interactive elements. Their organisation, taken as a whole, constitutes the breeding system of the species.

In such an organisation, heterosis due to over-dominance, additive effects of blocks of genes, level of ploidy and genome structure (auto or allopolyploidy), vegetative behaviour (annual or perennial), competitive ability at various stages of development, are all related to reproductive status; that is to say

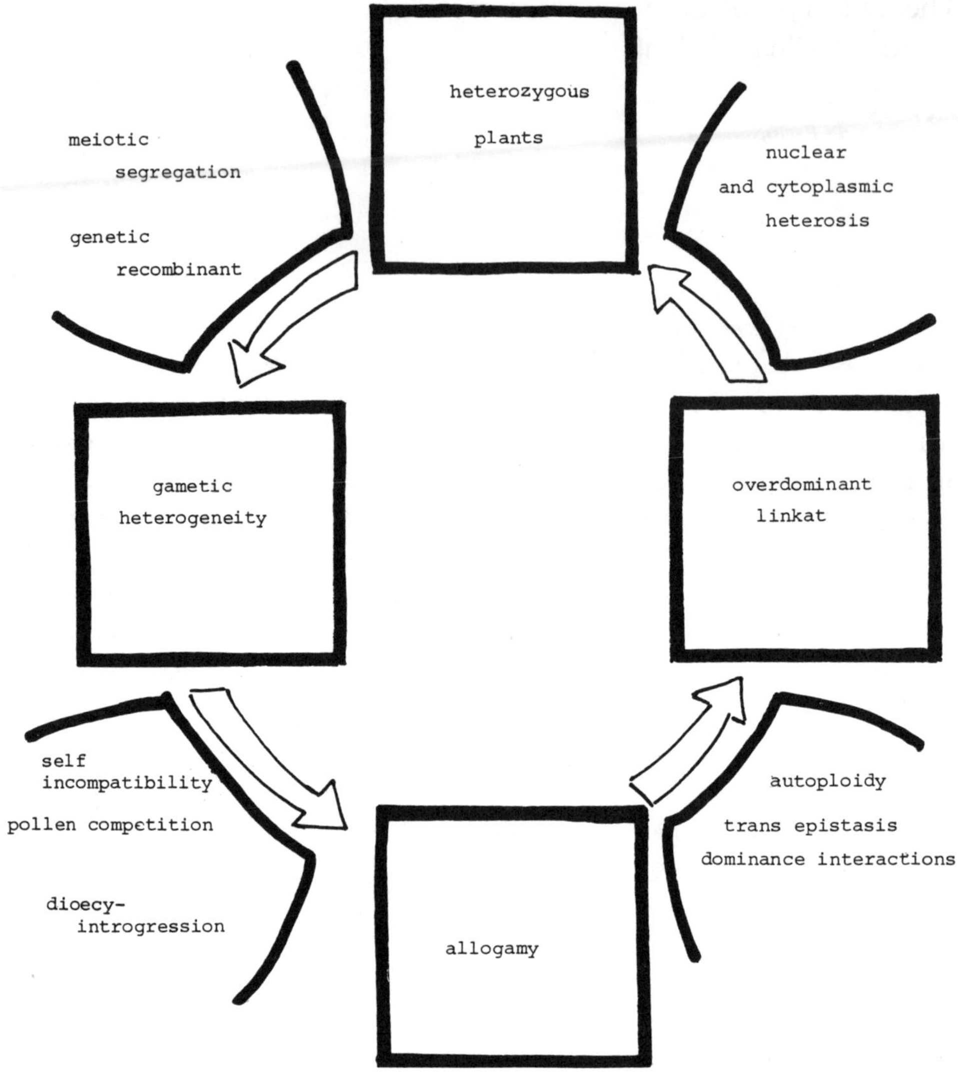

Fig. 1. The breeding system of a species.

total allogamy associated with self incompatibility, partial autogamy or total autogamy.

All these characteristics are interactive. They have been modelled simultaneously as the plant species differentiated in the course of evolution and they contribute to a kind of specific strategy (Figure 1). So each population of plants can react to the selection pressure of its natural environment.

For most species, this adaptative organisation is not fixed. On the contrary it includes alternative pathways which give a fairly good flexibility to the species. We know for instance the balance between vegetative and sexual reproduction in wild and cultivated potatoes, between suckering and out-crossing in raspberry, between tillering or apomictic seed or sexual seed in *Panicum maximum*, between outcrossing and inbreeding in lucerne. Such cases can be found in almost all plant species. Therefore, the rate of cross-pollination is not constant but it evolves according to environment, in natural as well as in cultivated conditions. Also because the breeding system acts as a whole, we can be sure that when a species tends to autogamy, or on the contrary, is improved by maximising allogamy, its gene expression, its gene arrangement, the frequency of its alleles will be changed for a better response to this new status.

So the important point to keep in mind when we intend to preserve the integrity of the characteristics of an ecotype or of a varietal population is that all the elements of the system have to be taken into account carefully.

At the genetic organisation level

In order to have a better understanding of the sensibility of the breeding system, let us look at the plant reaction at the more intimate level of its genetic system.

Our knowledge of the plant genome is rather limited. Most (80%) of the DNA in plants consists of repeated sequences which, we now think, play a major part in the quantitative regulation of gene expression. The coding genes are inserted in these repeated sequences with a large initiator promoter zone.

Generally the coding genes themselves are repeated. The repetition of each coding gene may be called redundancy, though each of these repeats may not be exactly the same and may show slight polymorphism. Inside each coding gene we know that there are fragments, some of them being translated by ribosomes, the exons, others being excised and remaining in the nucleus, the introns.

The exact enzymatic mechanism of excision and splicing at some precise site of the code is not yet known. However, as Gilbert (1978) implied, it may be that variability in these sites, in respect of age, history of the plant and environmental conditions, would offer a rich source of adaptative polymorphism.

So plant information possesses a flexibility through the possibility of combining variant arrangements between, or inside, blocks of coding elements: this constitutes the linkat theory which applies to various levels as follows:

- At genome level alloploidy allows an alternative function of homologous fragments: some of them play a 'leader' part as they belong to one 'leader' genome, the homologous sequences are vicariants, that is to say they act as a kind of modifier gene. But each time the leader genome becomes deficient, by mutation, by repression, or because regulatory conditions have changed, then, one of these vicariant genes plays the part of the major leader sequence.

The linkat has been proposed for the other two smaller levels of arrangement. In fact, these homologous fragments in the genomes contain not only one coding gene but are blocks of linked genes, protected from recombination by genes which inhibit or lower the number of crossings over.

The characters encoded in these blocks present a global adaptative value which is why evolution gathered them in such blocks.

However, these linkats are not definitively prevented from recombination, a low rate of crossing over can occur to give new linked code arrangements which in turn are submitted to natural selection. Attention must be drawn to the fact that a linkat is not a fixed stable unit; it is a protected semi-solid arrangement and in some cases, the chromosome itself can be considered as a super linkat.

- Another level of adaptative flexibility appears at the redundancy level of the sequential repeats of a gene. Each of them is more or less equivalent to one allele and there we can obtain an heterotic effect due to epistatic interactions between redundant-like codes.

Amongst these genes some are expressed, others are dumb or repressed. This situation is not definitively fixed; at another stage of development, or in another environment, the expressed part can be repressed and the dumb one be translated into proteins. A high degree of adaptability remains.

- The third level of flexibility consists of the variations in the excision-splicing mechanism, which may lead to variant expressions, starting from the entire prematured RNA (which is the total transcription of the gene) and giving many partial combinations between kinds of elementary bricks (which are the various exons following the number of excisions and the intensity of splicing). There lie probably the most important steps in variability of gene expression (Figure 2).

To this possibility of selection in the sites of excision and splicing we must add the fact that transposons or episomes, may be inserted inside the linkats and be a source of polymorphism in their expressions. A fairly good continuity seems to exist between viroses mitochondria episomes, transposing sequences, and exons-introns systems. All of them, either in a pathogenic way or in a normal way, add their own information to a convergent common phenotype.

The controlling factors of linkat expression

Therefore, the main question that arises is how a genic expression can be repeated in each individual of the same genotype, since there are so many possibilities of variation. To understand it, and to show how regeneration of cultivars (or of natural populations) has to take account of the fragility of interaction between genetic expression and environment, we must analyse the controlling procedures of the linkat.

To make a very long story short we can say that in every cell of the plant where genes are expressed, there are converging signals from correlation between organs, and from environmental

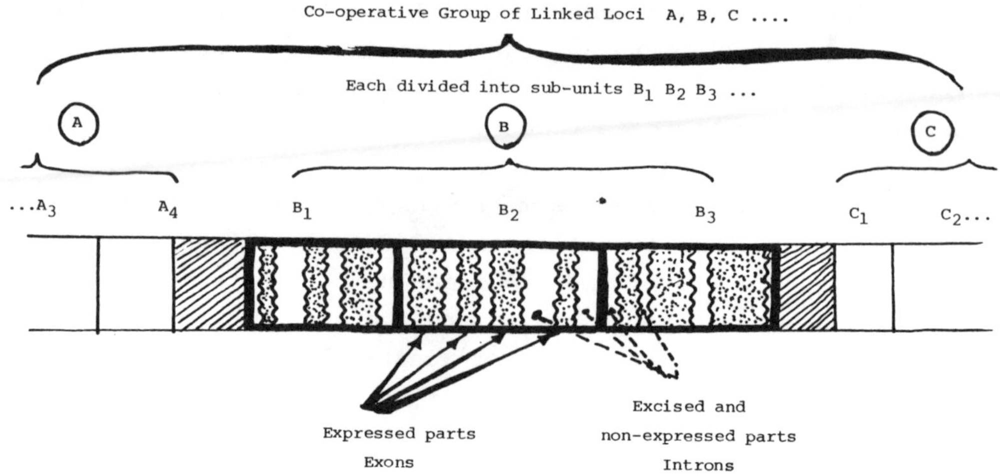

Fig. 2. The structure of a linkat.

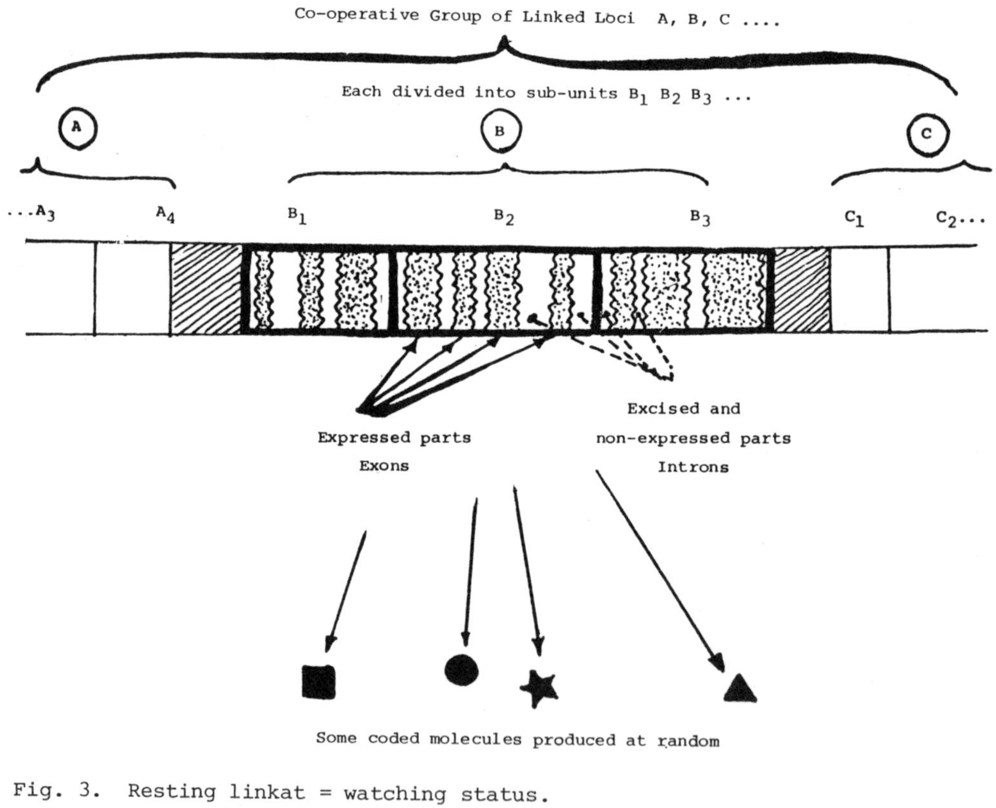

Fig. 3. Resting linkat = watching status.

conditions through leaves, root, and day/night physiological balances. These signals are translated into molecular situations, that is molecular interfering waves which determine a population of molecular shapes (that we may call an interferential substrate for the genes).

The various combinations of the molecules encoded in a linkat are produced at random at a very low frequency even in a resting genome. In some ways a linkat is never totally repressed. There is a kind of watching status of this genic system. If, at a given moment, the shape of the substrate population has a better affinity coefficient with some of the molecular arrangements coded by a linkat, then the activation of these best adapted combinations of genic units increases more and more to give a better gene response (Figures 3 and 4). In this hypothesis, the intensity of a genic combination derepression is related to its affinity with the molecular shapes in the cytoplasm, these latter being themselves dependent on all the cellular background and environmental signals.

It is easy to understand that gross environmental conditions and genetic structure of the plant determine the genic response of the plant. In such a view, the sensitivity of genes to environmental conditions is obvious. Of course, when a genotype is cultivated *in vitro*, its linkats are disorganised - this is the source of so many nuclear and cytoplasmic variants.

The sensitivity of the linkat to genic structure is probably not so evident. Yet the numerous epistatic effects play an important part in its flexibility of expression. We may speculate that, if in an inbreeding system, an allogamous species is disturbed in its trans-epistatic cohesion, this will yield a poor genic expression. At the same time, autogamous species which are outcrossed can also be disturbed, because most of the organising strength of their linkats belong to cis-epistacy (Figure 5).

These hypotheses are in agreement with experimental facts which show that not only the allelic frequencies but also the modelling of allelic expression, are under the control of natural selection.

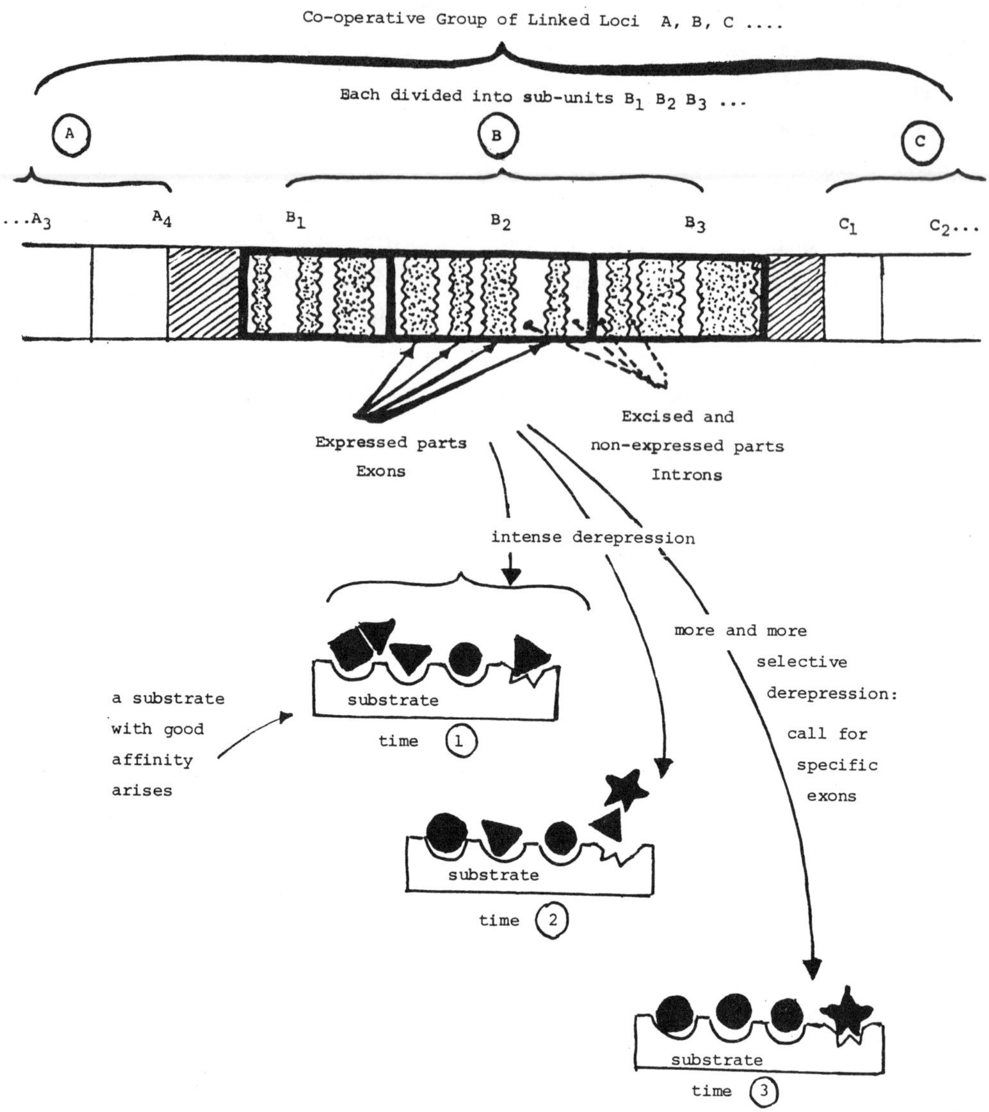

Fig. 4. Control of a linkat.

EFFECTS DUE TO TRANSFER OF AN ALLOGAMOUS POPULATION FROM ONE ECOSYSTEM TO ANOTHER SITE

Sampling problems

It is very difficult to sample and to transfer a sample from one area to another.

Suppose we take a sample of N unrelated plants from an allogamous population or synthetic variety: if we maintain the effective size (N) of the sample constant in a collection at each multiplication, one plant in the nth generation will possess at most 2^n ancestors in the initial generation. As there are only N potentially different ancestors, the probability of matings between plants originating from the same ancestor increases at each generation.

Hence the formula which gives the increase of inbreeding coefficient F with the generation of multiplication for various effectives of the samples.

$$F_n \simeq 1 - (1 - 1/2N)^n$$

The risk of inbreeding is large as early as the first multiplication for small samples ($N < 10$). For larger samples ($50 < N < 100$) it becomes serious after five generations.

The Hardy Weinberg law of stability for genic frequencies applies only if the effective population size is large. In small populations genic frequencies show random variations due to bias in gametic arrays. The frequency of an allele varies from 0 to 1 following a probability low. Namely its variance is given by:

$$V = p_o \ q_o \ (1 - (1 - 1/2N)^n)$$

(if p_o is the frequency for an allele at a locus and q_o the frequency for the other alleles).

Therefore the dispersion of genic frequencies increases at each new generation. Each time it randomly reaches 0 or 1, allelic variability is definitively lost. This is known as genetic drift.

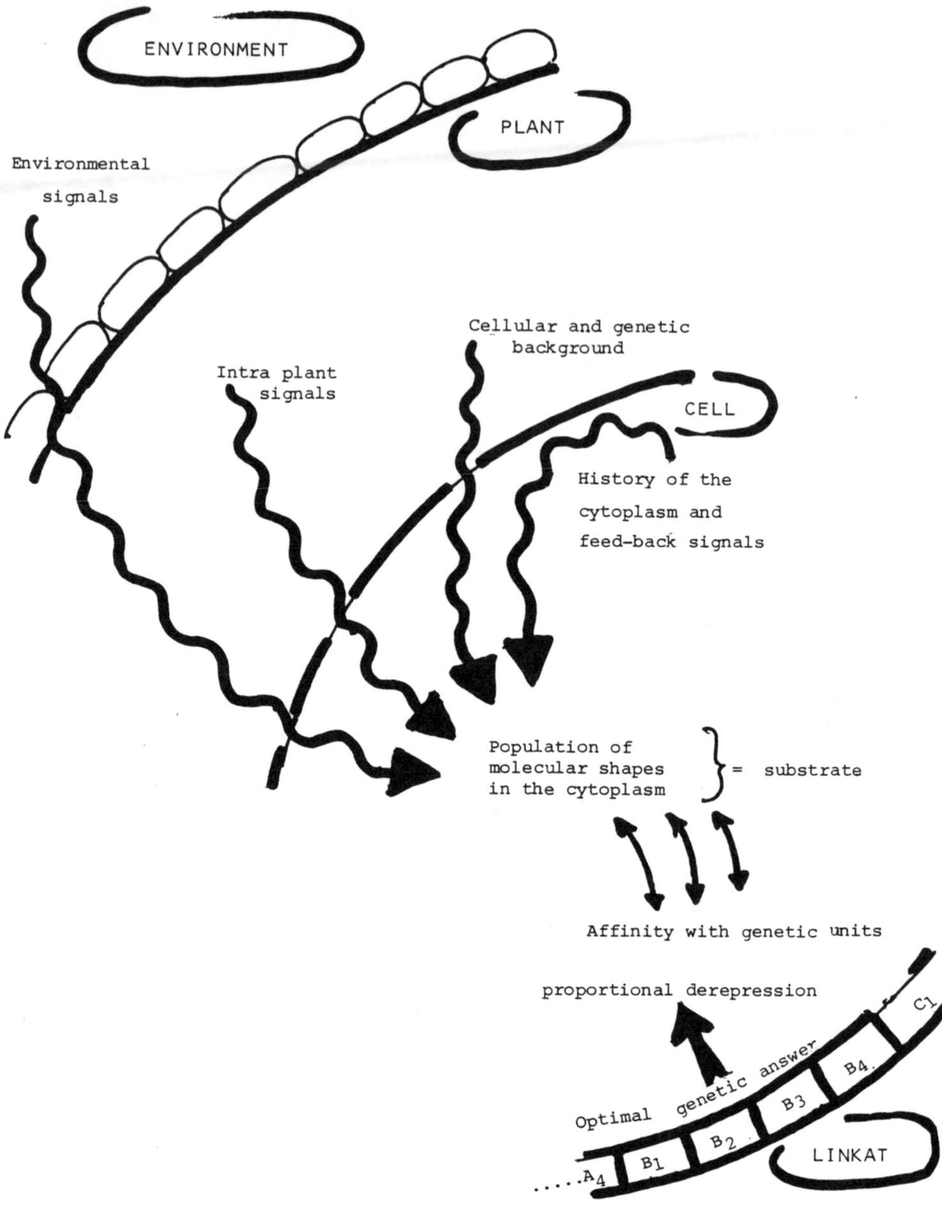

Fig. 5. Controlling signals for a linkat.

Directional selective pressure

If we assume that the sample was collected in a representative area, then the sampled population is distributed normally. In the new area (for instance in a genebank centre where it is grown afterwards, the sample is submitted to new environmental effects, only one part of the population is kept at each generation, and this is a case of directional selection. These effects will affect both mean and variance of the sample.

For a given trait, after one generation we observe a response R to the directional pressure, which is the deviation of the mean value.

$$R = S\ h^2$$

where S is the mean value of the parents allowed to produce progeny in the new conditions and h^2 is the heritability of the trait.

If we introduce the intensity of selection $i = \frac{S}{\sigma_p}$ which is the standardised value of selected parents we get:

$$R = ih^2\ \sigma_p$$

The variance σ^2p will be reduced following the expression:

$$\sigma^2p\ (1 - i(i - x))$$

where x is the abciss value of selected parents in the normal distribution.

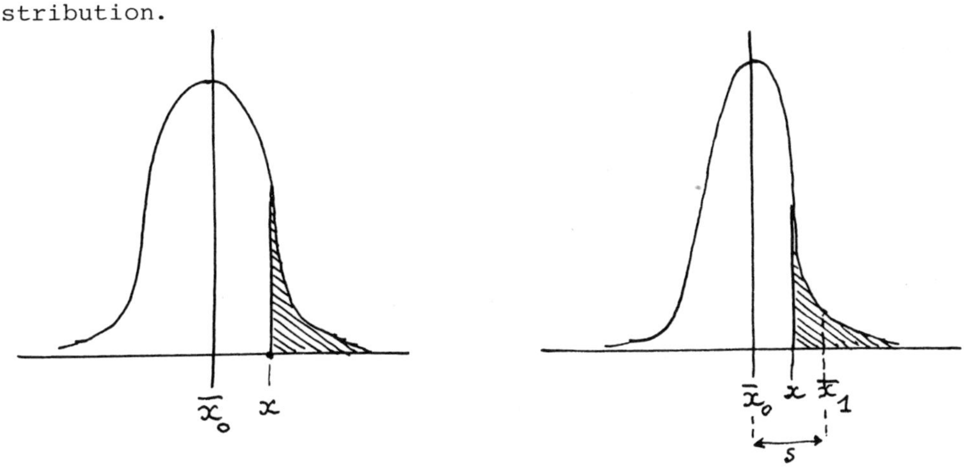

$\bar{x}_o$ = mean of total population

x = inferior limit of selected parents

$\bar{x}_1$ = mean of selected population

s = $\bar{x}_1 - \bar{x}_o$

Fig. 6.

This deviation of mean value and reduction of variance will be more and more drastic as the number of generations under the new selection pressure increases. At the same time we observe a gametic and a zygotic drift.

One of the most clear-cut examples is the change due to day length sensitivity in the vegetal flowering process. A slight change in day length conditions will prevent a certain number of plants from flowering and the above mentioned formulae will apply.

The case of pollination

For most species cross-pollinated by insects the population of pollinators have specific habits. For example, a field pollinated by domestic bees does not show the same amount of intra-individual pollination (selfings) as if it were pollinated by bumble bees. This would mean that when transferring a plant population we should also have to transfer its original population of pollinators. We know that during the process which led from wild species to domesticated cultivars, the absence of pollinators was one of the various sources of the change from cross-pollination to autogamy. This might appear to be a relatively unimportant factor, but it has drastic consequences on the progeny, when after transfer to new conditions the population finds inaccurate pollinators.

For wind pollinated species, the trouble can arise from a change in weather conditions. We know that pollen shedding is dependent on temperature, humidity, day length, etc. Under other climatic environments it could be reduced and self-pollination would increase abnormally. For many reasons the integrity of a heterogeneous population of plants is not as easy to maintain as one might imagine.

The species-complexes

A species rarely constitutes an isolated reproductive unit.

Crosses between wild and cultivated races of domesticated species are frequent. They produce those kinds of weeds which are subordinated to man's agriculture and evolve together with cultivars. So there are evolutive groups, or species-complexes consisting of wild forms, weeds, and cultivated varieties which are submitted to coevolution by permanent exchange of genic flow.

This phenomenon is not so common in our countries where the purity of variety is under a strict control. Nevertheless most of our grasses, or clover or lucerne varieties, are polluted by 'weeds' or wild forms. More frequent, and probably with positive effects, are the examples in tropical crops such as rice, sorghum, millet etc. In some cases weed-forms induce variability and fitness to cultivated forms. In other cases, they play their part as competitors in eliminating the weaker phenotypes. These weeds escape until grain harvesting because of brittle rachis (they resow spontaneously) but they contribute as pollinators to the genetic pool. When a seed sample is collected in the field it cannot include the specific weed population, seeds of which have already dropped off. After two or three generations of multiplication of this kind of sample, the absence of gene flow from weed pollinators, and of competitive selection pressure cause a bias in the reproduced cultivar. At the same time as a culture is transferred from one place to another, the many parasites, which maintained a continuous specific pressure in the initial conditions, will disappear. The collected sample will no longer be subjected to their differential selection effects and will evolve in another direction.

These remarks stress again the limits of collections reproduced without the pressures which maintain a high genetic load.

Control of the environment in agricultural systems

One of the most striking features of agricultural systems is that the environment is increasingly under man's control.

- Biological environment

Effects of density, uniformity of seeds, uniformity of age, weed control, control of pathogens, inoculation, premunition, mycorhization, etc.

- Chemical environment

Fertilisers, manure, special oligoelements, rate of CO_2 in greenhouses, hydroponics, etc.

- Physical environment

Adjustment of water requirements, drainage and irrigation, freeze control in vines, fruit, etc., air and soil temperature control, light and moisture in greenhouses, soil temperature in fields with black plastic, etc.

The object of these management techniques is to optimise crop production from an economic point of view. Such agricultural management is related to seed preservation in that such systems require and preserve strictly adapted genetic structure of cultivars. There is a negative correlation between the sophisticated control system in the most intensive agricultural systems and the genetic diversity of the plant. This is a consequence of the resonance law of load theory. The environment is less predictable, its environmental load is higher and in response the genetic load of crops has to be higher. On the other hand, sophisticated agriculture requires specialised cultivars such as single hybrid corn or tomato hybrids.

For seed preservation the conclusion would be that it is not only necessary to preserve the genetic structure but the agricultural system itself with its specific load.

Conclusions

It is usually assumed that plants possess a high degree of flexibility to respond to environmental changes. We have shown that both at the individual and population levels this plasticity is linked to definitive genetic modifications. As conventional genetics points out, this change involves not only allelic frequencies but reproductive patterns and linkat stability: consequently genic expression itself. So it is a deeper modification of the breeding system of a species. There lies the main difficulty in preserving genetic resources.

POSSIBLE WAYS

Various ways can be tested to keep the integrity of an ecotype or of a cultivar in cross-breeding species. We shall briefly review each of them, from the more natural, to the more sophisticated ones (Table 1).

TABLE 1

POSSIBLE WAYS

- *In situ* preservation	- Who is able to maintain an ecosystem?
- Living collections	- Original integrity (whole genotype? one gene? cytoplasm?) or Recurrent selection.
- Seed storage	- Periodically renewed?
- Sophisticated methods	- Keep cellular strains in cold. Keep unfertilised eggs, or pollen. Keep DNA only, inserted in bacterial strains.

In situ preservation

Most of the theoretical remarks show that it is difficult to maintain genetic equilibrium: allelic ratio, linkat stability, permanent expression of gene, composition of cytoplasms in terms of their populations of organites. It is clear that only by preserving the whole breeding system (and this means the whole ecosystem) shall we be able to maintain the genetic wealth of an ecotype. So the solution would be to determine, for our most important crops, which natural populations, which ecotypes, which cultivars have to be preserved, and having done so, to determine where the optimal ecosystem for each of them is located. The second phase would be to determine what course to pursue in order to keep these specific ecosystems which contain the required ecotype or cultivar in small isolates.

The questions raised by this proposal are numerous and the answers very limited. The most important are as follows:

- Do we need to maintain all the existing genetic diversity?
- If not, what would be the criteria for a choice?
- How is it possible to organise plant reserves *in situ*? If we mean cultivated plant reserves, we have to try to maintain an agricultural ecosystem without any biological or technological evolution.

- Is this possible? Who will pay for such farms specialising in conservation agriculture?
- Is it realistic to propose to preserve a given level of artificial selection pressure? Who will be able to ensure that populations of pests or pathogens will not change?

In my opinion, this way must be attempted for very important crops. It will take a long time to set up such an organisation and it will be very expensive. However, it might be a good proposal to try it on two major crops as models. I would suggest lucerne because western Europe is probably one of the richest crossways for this plant. I would also propose maize. Though our countries are only secondary diversification centres, a real and long-standing European diversity exists.

For perennials or tree crops the problems are of another nature because without any genetic pollution we can maintain the integrity of the individual composition of an ecotype over many years. However, older individuals will die and we shall have to maintain their progenitor role in the preserved population. If vegetative propagation is not possible the same problems arise as are listed above.

Living collections

By the organisation of collection centres we intend to gather many species and many cultivars in one experimental place. The main point about this system is that we know that the normal selection pressures will be disturbed by the transfer of cultivars and ecotypes from original conditions to a collection centre.

In this kind of preservation, we have to discuss carefully the objectives: do we want to keep the original genetic integrity of a sample at any cost, or do we want, as most breeders do, to ensure genetic progress by improving 'back-up' population?

If we choose the original integrity then a demographic study of each specific case is needed. But are we confident that each year we shall be able to go back to the original source of material? If not, because of agricultural, economic, or political evolution, we must raise the following question:

Can we control and minimise the risks of genetic drifts? First of all is the evident necessity to isolate each sample from the other. This is expensive if many cultivars are gathered at the same place. Then we have to be sure that the environmental conditions in the collection fields do not prevent panmixia (or the normally weighted outcrossing of the species). In most cases, it seems important to plan for artificial cross-breeding; hand crossing such as artificial polycrosses, or systematic pollination. A tight control is obvious too: the plant density has to be correct and the sanitary control good, in order to avoid plant selection during development. The living collections must be carefully protected against any limiting factors from the environment unless these limiting factors are also present in the area from which they come. So a high degree of protection against abnormal frost, drought, heat, pathogenic attack (and so on) has to be organised in such field collections.

On the other hand, if the aim was to back up original samples then various methods could be used. Most of them belong to recurrent selection schemes. At the lower cost the choice of progenitors would be made by phenotypic mass selection, and only promising phenotypes would be allowed to cross. A valuable cheap system could be the chain-cross method. It consists of spaced planting of the plants in rows, to eliminate all those which do not look satisfactory during vegetative growth. Then, just before flowering, the remaining plants are numbered from 1 to n and crosses are organised as follows: number 1 will be used to pollinate 2, number 2 to pollinate 3, ... number n - 1 to pollinate n, number n to pollinate 1. There is a kind of chain of crosses. If during seed formation, one plant seems not to be valuable it is discarded and the following number, which received its pollen is discarded too. Then all remaining plants will be harvested together.

A more time-consuming method might be to extract from each sample a rough synthetic variety, more or less representative of this sample. This method could be improved by repeating the multiplications of such synthetics in several places and by mixture and exchange, at each stage of multiplication, of the seeds collected at every place. So if there are directional selection pressures in one place they will be balanced by the seed mixture from other places. This method could be named 'tempered synthetics'.

Seed storage

If we could store seeds in good conditions without any drift due to differential germinative abilities with genotypes, seed storage would probably be the best way to keep the integrity of a sample. But in polymorphic populations we know many examples of differences in seed longevity related to other characteristics under genetic control (seed envelope colour, shape and size of seeds, permeability of teguments, etc.).

In these conditions we intend to preserve an ecotypic variability by storage of seed samples in the best conditions: storage in sealed cans with control of low seed moisture with absorbants, control of atmosphere (inert gas or vacuum), control of low temperature in conditioned chambers.

We would expect when we sow one of these samples to find again the genetic diversity. This may be true for a 5, 10 years storage but progressively some genotypes will selectively lose their germinability. Nevertheless such a method could save time and space if it was included in a system where seeds are periodically renewed by one generation to produce seeds in the field. There we encounter again the difficulty of reproducing a genetic polymorphism.

More sophisticated methods

The recent developments in plant regeneration from cells or from gametophytes could be a way to store each individual of a population under the form of its cells. This kind of experiment has been conducted in several places over the last few years. It is difficult to know what results they will give because of the rapid improvement of the techniques. However, we already know some of the constraints of these methods.

Imagine that we establish cellular strains representing the major amount of variability in an ecotype. These strains have to be transplanted from tubes or dishes into new media at a specific frequency (1, 2, 3 months). This is tedious work, even in a routine laboratory. Furthermore if these strains are stabilised in media where they do not regenerate plants, we have to induce the production of plantlets when we need to return to the plant population. It is well known that the frequency of strain transplantation will induce various modifications in age and genetic programme of the newly formed plantlets. Between each cellular division the age (measured as total correlation signals received) in the epigenic structures of the cell will progress. So the regenerated plants after 1 month are different from regenerated plants after 6 months. It is not infrequent for strains progressively to lose their ability to express morphogenesis. During this evolution regenerated plants are more and more variants and mutants. These remarks indicate that storage of these strains at a low temperature, say 4°C, where metabolism is low, is not a good solution because the greater the rate of mitosis, the slower will be the evolution of the genetic programme (i.e. age).

Having observed such complexities, some laboratories are now trying to stop the strains. Various techniques have been tried such as liquid nitrogen storage, anaerobic storage and lyophilisation. It seems that deep frost storage, though expensive, could be an efficient way. If the techniques of cell regeneration are to progress in the coming years, then an ecotype could be represented by cellular or embryoid populations congealed in small dishes or tubes. At the moment only a few species allow cell cultures followed by embryoid production and plant regeneration.

Another method could develop - gametophytic storage. It is possible for some species to store mature pollen; on the other hand, it is becoming possible to cultivate unfertilised ovules. As such techniques are improved a very satisfactory solution might be liquid nitrogen storage of both male and female gametophytes followed by *in vitro* fertilisation. Such research must be encouraged.

Gametophyte cultures *in vitro* allow the production of haploid plants: (artificial androgenesis and, more recently, artificial gynogenesis). In both cases immature gametic cells are needed. The storage in liquid nitrogen of immature microspores of rice followed by haploid plant regeneration has been demonstrated. Though pollen plants can be doubled to diploid they do not represent the population of origin exactly. Many studies have shown that a large variation appears. However, for collecting large wild populations, the storage of immature microspores or macrospores could be one new solution in the future. But many variants, genetic drift, genotypic competition must be expected with these techniques.

The most futuristic way would be neither to store plants, nor to store cells, but to try to store DNA information only. Theoretically it is possible to cut DNA molecules, with restriction enzymes, and to insert pieces of the plant DNA in a collection of bacterial strains. The actual problems lie in identification of pieces (probes are difficult to prepare) and in reinsertion of the DNA fragments into a plant which will be used as genitor of this information

CONCLUSION: THE NEED FOR INTEGRATED METHODS

Our discussion centres upon preservation and storage of genetic diversity. Another connected problem should be analysed: how to name and describe each entry.

It is a wide-ranging and difficult problem for allogamous plant populations. To stress the most important point we can say that every characteristic must be described with its own mean and genetic variance whether it is a morphological trait or an agronomic performance. We have not discussed this matter here but it needs serious development.

Coming back to the subject, one basic question is whether we have to preserve an exact pattern of the actual diversity of a cross-pollinated species or only keep the alleles or the genic arrangements of a linkat. In other words, is it the individual which is useful for breeders or some special gene or genetic combination?

The answer is not so clear. Some of us will say we need genitors to transfer polygenic traits, others will speak in terms of chromosomes or of linkats composition, others will be aware of cytoplasms and gene-cytoplasm interactions. It seems that we do not know the structure and regulation of plant information sufficiently well to say whether we will be able to rebuild all the genetic polymorphism of a species from its pieces of DNA.

This would suggest using several methods simultaneously, each of them with its own qualities and defects. The organisation of this necessary task needs money, place, time and co-ordination. A constructive proposal would probably be to start with one or two important crops as models and in these crops to choose a number of types to save. The general scheme might be as follows:

1) To determine a few areas where the model plant grows and expresses natural polymorphism. Each of these limited areas could be considered as natural reserves for one group of ecotypes or cultivars having some affinity or relationship.

2) At the same time to put in collection fields or collection centres the same range of cultivars and to try to reproduce them, either by chain crosses or tempered synthetics, or simply by natural isolated plots.

3) It seems necessary to use seed storage simultaneously with two subtreatments: long term storage, compared with interrupted storage (one plant generation to reproduce seeds at given periods).

4) To choose a tissue culture laboratory where operations such as cell cultures and gametic cultures are well known and to try with it an assay of cellular storage - on a reasonably limited scale.

5) To nominate a co-ordinating board which will distribute and follow the operations and organise at given periods (for instance each five or ten years) field comparisons of plants from every method. I know that various research projects have already begun in this area in several countries. The task is perhaps only to focus them on the same model crops and on representative types used as standards in international experimental designs (Figure 7).

To end with an optimistic view I would say that every day thousands of plant breeders bring a large new genetic diversity.

- CHOOSE ONE OR TWO MODEL SPECIES
- ORGANISE INTERNATIONAL CO-OPERATION

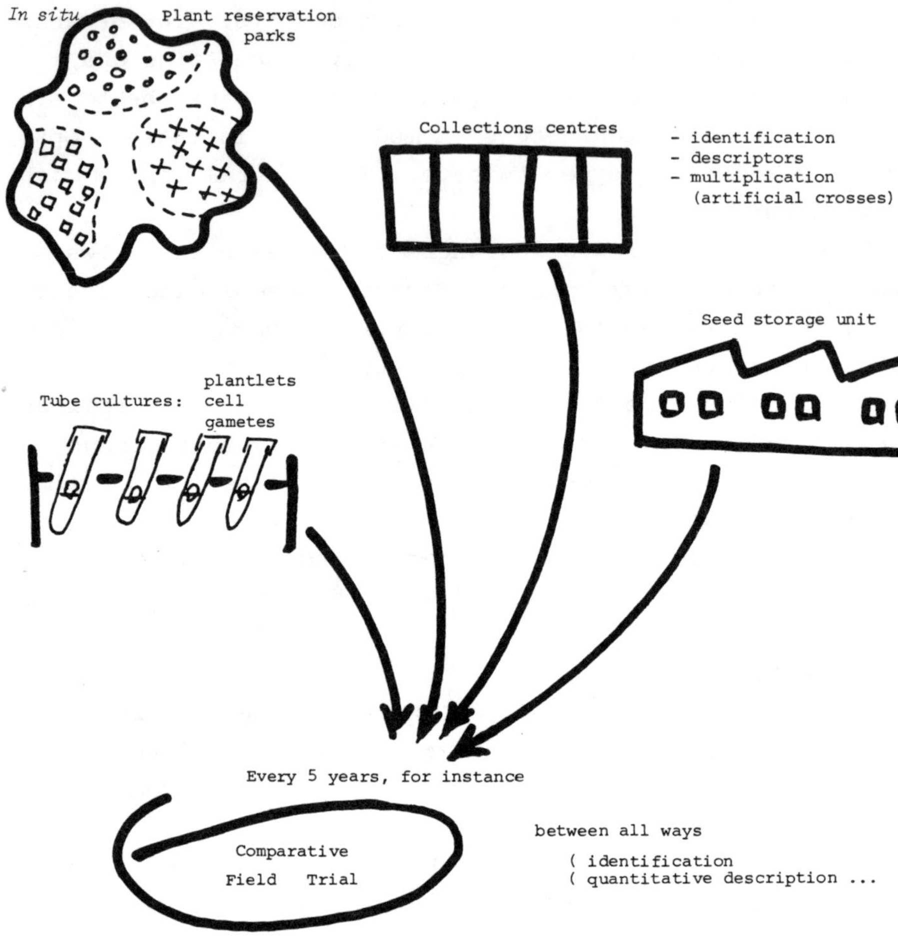

Fig. 7. Proposal = Experiment on integrated methods.

The reason for preserving genetic resources is not because we are losing this diversity but because we are aware that it is a wealth for future improvements.

REFERENCES

Ahmadi, N., 1979. Etude biométrie de la variabilité créée par trois voies de reproduction autofécondation, androgenèse et gynogenèse) chez *Hordeum vulgare* L. Diplôme d'Etudes Approfondies. Amélioration des Plantes. Université Paris-Sud-Orsay.

Andre-Truong, I., 1977. Variabilité des plantes issues d'androgenèse *in vitro* et tentative d'application directe en sélection de cette variabilité et de la mutagenèse par voie haploide chez le Riz *(Oryza sativa* L.). Thèse de 3è Cycle. Amélioration des Plantes. Université Paris-Sud-Orsay.

Belliard, G. and Pelletier, G., 1979. Protoplast fusion as a tool for cytoplasmic genetics in higher plants. Behaviour of debneyi and tabacum chloroplasts and mitochondria in their cytoplasmic parasexual hybrids. Vè Congrès International sur les protoplasts, Szeged (Hongrie), 1979.

Cauderon, A., 1980. Sur la protection des ressources génétiques en relation avec leur surveillance, leur modelage et leur utilisation. Extrait du Procès-Verbal de la séance du ler octobre 1980 à l'Académie d'Agriculture de France.

Davies, D.R. and Hopwood, D.A., 1980. The Plant Genome. 273 p. ed. John Innes Charity, Norwich.

Demarly, Y., 1977. Génétique et Amélioration des Plantes. 287 p. ed. Masson, Paris.

Demarly, Y., 1978. The concept of Linkat. Conf. Broadening Genet. Base Crops, Wageningen. ed. Pudoc, Wageningen, 257-265.

Franklin and Lewontin, R.G., 1967. Is the gene the unit of selection? Genetics, 1967, 65, 707-734.

Gilbert, W., 1978. Why genes in pieces? Nature, 1978. 271, 501.

Griffing, B., 1960. Theoretical consequences of truncation selection based on the individual phenotype. Aust. J. Biol. Sci., 13, 307-343.

Kimura, M., 1958. On the change of population fitness by natural selection. Heredity, 12, 145-167.

Kimura, M., 1968. Evolutionary rate at the molecular level. Nature, 217, 624-626.

Kimura, M. and Crow, J.F., 1964. The number of alleles that can be maintained in a finite population. Genetics, 49, 725-738.

Kouadio, Y., 1979. Analyse génétique de la variabilité observée dans la descendance d'une lignée pure de Riz *(Oryza sativa* L. var. Cigalon) traité par androgenèse *in vitro*. Thèse de Docteur-Ingénieur, Université Paris-Sud, Orsay.

Ollivier, L., 1973. Le calcul de l'effectif génétique des populations animales. Ann. Génét. Sél. anim (INRA), 5, 363-368.

Ollivier, L., 1974a. La régression parent-descendant dans le cas de descendances subdivisées en familles de taille inégale. Biometrics, 30, 59-66.

Ollivier, L., 1981. Eléments de Génétique Quantitative. INRA Actualités Scientifiques et Agronomiques, 152 p., ed. Masson, Paris.

Pernes, J., Savidan, Y. and Rene-Chaume, R., 1975. *Panicum*: structures génétiques du complexe des 'Maximae' et organisation de ses populations naturelles en relation avec la spéciation. Boissiera, 24, 383-402.

Prings, D.R. and Leving, C.S., 1978. Heterogeneity of maize cytoplasmic genomes among male-sterile cytoplasms. Genetics, 89, 121-136.
Quetier, F. and Vedel, F., 1977. Heterogenous population of mitochondrial DNA molecules in higher plants. Nature, 268, No. 5618, 365-368.
San Noeum, L.H., 1976. Haploides d'*Hordeum vulgare* L. par culture *in vitro* d'ovaires non fécondés. Ann. Amélior. Plantes, 26, 751-754.
Sibi, M., 1978. Expression of cryptic genetic factors *in vivo* and *in vitro*. Proc. Conf. Broadening Genet. Base Crops, Wageningen. Pudoc Wageningen, 339-340.
Waddington, C.H., 1957. The strategy of the genes. G. Allen et Unwin Ltd., London, 262 p.
Wallace, B., 1963. Modes of reproduction and their genetic consequences in Statistical Genetics and Plant Breeding p. 3-21. NAS/NRC Publication 982, Washington DC.
Zuckerkandl, E., 1975. The appearance of new structures and functions in proteins during evolution. J. Mol. Evol. 7, 1-57.

DISCUSSION

W. Plarre *(Federal Republic of Germany)* Concerning your recommendations, you are very sceptical regarding preservation *in situ*. I think we must consider this. I have made a proposal as a contribution to this meeting, perhaps on Friday we can discuss this. If we are convinced that it is necessary to maintain our out-pollinated crops, especially the land races, then we can only do this if we protect the whole environment from which they are derived. Let me give one example. There is a variety or rye, an old variety cultivated under bad conditions in which there is late frost. This variety is resistant to late frosts and, therefore, we need an environment in which there are always late frosts in order to maintain the genotypes with this frost resistance.

Y. Demarly *(France)* Yes, I agree.

J.T. Williams *(FAO)* In relation to *in situ* conservation, it may be of interest to let people know of some recent developments. A study of this problem has been made by IUCN on behalf of the International Board for Plant Genetic Resources (IBPGR), as to whether or not it is feasible to preserve environments, particularly for land races and wild relatives of cultivated plants. The general consensus was that it is not really feasible to preserve the primitive agricultural systems but, of course, this is taking place in certain parts of the world - there are reserves designated in Himalayas and in Sri Lanka for certain types of plants. What has been proposed is that a working group or a co-ordinating committee should be set up to look at this whole grey area between programmes of UNESCO, FAO, IUCN, UNEP, IBPGR. I do not know quite how this is going to be organised or when it is going to be set up. I suspect that it will take some time and there will be a lack of funds for it. However, I think there is a lot of interest in this problem. We have also heard of various attempts to try to preserve some of the more primitive agricultural systems, particularly in relation to pasture plants in parts of eastern Europe. I am not sure whether this has been done on a large scale as yet, but there is an international initiative and I think it would be interesting if this meeting finally comes to some recommendations to support it.

Y. Demarly I agree but the basic problem is to keep many cultivars and many ecotypes and not just a few which are exceptional for one characteristic. Professor Plarre's example for rye is very interesting, but to keep all rye diversity by *in situ* parks is very difficult.

J.G. Hawkes *(UK)* Yes, I am sure you are right when you say that various methods need to be used; there is not one single method that we can adopt.

Taxonomic aspects of the germplasm conservation of cross-pollinated cultivated plants

W.A.BRANDENBURG, E.H.OOST & J.G.VAN DE VOOREN
Agricultural University Wageningen, Netherlands

ABSTRACT

In Genetic Resources Centres (GRCs) documentation about accessions is standardised by means of descriptors, e.g. place of collection, morphological features, agricultural value. Botanical and/or fancy names are also considered to be descriptors.

However, there is one essential condition for the usefulness of names as descriptors, namely the check on whether names are correct. If not regulated by the International Code of Botanical Nomenclature, 1978 (ICBN, 1978), the nomenclature of cultivated plants is directed by the International Code of Nomenclature for Cultivated Plants, 1980 (ICNCP, 1980).

For cultivars, including land races, however, ICNCP, 1980, does not contain rules concerning any typification. So, in these cases there is no well-defined connection between a plant material and its name. Particularly where it concerns the variation and the need for stability of cultivars of cross-pollinated plants, the lack of this connection is even more striking.

The rapidly increasing amount of accessions of cultivated plants in GRCs is one of the main reasons for focusing attention on this problem.

This paper proposes a general procedure for checking cultivar names, based on living and conserved standards, descriptions and memory. The possible incorporation of the proposed procedure into a list of descriptors is discussed.

NOMENCLATURE

The International Code of Botanical Nomenclature, 1978 (ICBN, 1978) regulates the nomenclature of plants by principles, rules and recommendations. For each rank, the notation of the Latin name of the corresponding taxon is ruled, e.g. by a fixed suffix. This means that it can be derived from the name, whether the taxon is a genus, a species or a botanical variety. Before a name is legitimised, the following conditions have to be fulfilled:

- Latin diagnosis and description;
- valid publication;
- designation of a type.

By the principle of priority the ICBN governs which of the legitimate names is correct for the same taxon. In cases of homonymy, the same principle guides the decision as to how a legitimate name has to be correctly applied.

By the type method, the connection between plant material and its name is made. The type of a genus is a species which is designated as such by the author. In other words, the type species always belongs to the genus to which it is assigned. In the case of a species or a taxon of lower rank, the type is a well conserved specimen, or, when this is not available, an illustration. Type, diagnosis, and full description, illustrations included, form the protologue, the starting point for a definitive identification of plant material. If not regulated by the ICBN, the nomenclature of cultivated plants is directed by the International Code of Nomenclature for Cultivated Plants, 1980 (ICNCP, 1980). In the latter Code, the cultivar (cultivated variety) has been defined in Article 10 as follows:

> "The international term cultivar denotes an assemblage of cultivated plants which is clearly distinguished by any characters (morphological, physiological, cytological, chemical, or others) and which, when reproduced (sexually or asexually) retains its distinguishing characters."

Dependent on mode of reproduction, the following categories of cultivars can be distinguished:

- Vegetatively propagated
  - clone
    multiclone, mixture of closely resembling clones
- Generatively propagated
  - line
    multiline, mixture of closely resembling lines
  - F_1 hybrid
  - entity of cross-pollinated plants

Comparing both definitions of cultivar and land race, it should be apparent that according to the ICNCP a land race has to be considered as a cultivar (Harlan, 1975). Consequently, the nomenclature of a land race is also directed by the ICNCP:

- After 1st January, 1959 only fancy names within brackets or after the abbreviation cv. are admitted. Latin names, given before this date and derived from botanical varieties, are also admissible. The cultivar name is placed after the botanical name of the taxon with the lowest possible rank.
- Translated synonyms in other common languages than the original are allowed.
- Later homonyms, generally accepted and widely used, are preferred to original names, which are afterwards rediscovered.

As to the cultivars, the ICNCP does not rule any typification. In order to be legitimate, the ICNCP states that a cultivar name has to be validly published. Names given before 1st January, 1959 which have not been validly published, can be legitimised by a registration authority. The publication and registration list must contain a description of the cultivar.

Cultivars are often adapted to a small range of environments. The expression especially of quantitative characters is greatly influenced by locality and time. Although quantitative characters (for example yield and earliness) are often of great economic importance, they have a restricted descriptive value for the above mentioned reason. For an adequate description it is useful to look for qualitative characters, which do not necessarily have to be of economic importance.

Cultivars do not form a closed classification as in botanical classification under the ICBN (Figure 1). Between cultivars of the same botanical taxon certain gaps, containing non-cultivated plants, can exist (Figure 2). Consequently, not only description, but circumscription, based on any representative element, should be considered essential.

In connecting a cultivar with its name, the type method in the conception of the ICBN cannot be applied for the following reasons:

- A type does not necessarily have to be a representative element.
- A type of any living plant material cannot exist.

Fig. 1. Closed classification under the ICBN. Hierarchical order of ranks.

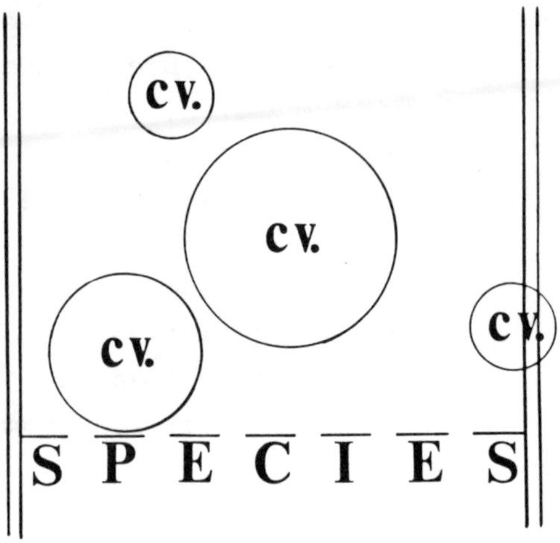

Fig. 2. Open classification of cultivars, independent of botanical classification.

Instead of a type, we strongly recommend using a living and conserved standard, both consisting of a representative sample of the cultivar. The mean and variation of characters within a cultivar are fixed by these standards. A living standard is important, because of its independence of locality and time. Being sampled once, a conserved standard is determined by the then prevalent environmental factors. Therefore, it is less useful than the living standard. However, it is valuable for a longer term, in case the living standard is absent or lost. By using both standards, a reliable connection between the cultivar and its name can be made.

TAXONOMY AND GERMPLASM OF CROSS-POLLINATED PLANTS

The aim of Genetic Resources Centres (GRCs) is to conserve as wide as possible a range of genetic diversity, with potential agricultural value. At present, among all the accessions of GRCs, cultivars, including land races, are the most frequent items (Ayad et al., 1980).

There are two main reasons for conserving germplasm by means of cultivars. Firstly, cultivars representing certain combinations of characters are entities that have already proved their usefulness in agriculture. So, asking for cultivars, plant breeders will find the desired genes in a more or less balanced genetic background, formed under the process of domestication and breeding. Secondly, a cultivar is indicated by a name. This name forms an entry in all documentation about the cultivar. In the case of old cultivars, one can trace back original descriptions or illustrations and even make use of oral tradition. The documentation on cultivars acquired in this way can be used for definitive identification of an accession. Likewise, if additional knowledge about origin, special characters and (historical) use etc. is obtained, this can be of great importance to plant breeders and crop evolutionists.

In cross-pollinated plants, there are obvious advantages in conserving especially old cultivars, provided that they can be obtained in sufficiently large quantities of seeds. Because these cultivars are sufficiently heterogeneous, thus securing conservation of genetic diversity, the number of accessions can be restricted. Consequently, more time per accession becomes available to complete the documentation of each accession.

PROCEDURES TO CHECK THE IDENTIFICATION OF ACCESSIONS

In GRCs, documentation on accessions is standardised by means of descriptors, e.g. place of collection, morphological features, agricultural value (Seidewitz, 1973, 1979). Botanical names as well as cultivar names are also considered to be descriptors, because they supply intrinsic information about the plant material concerned.

There is, however, one essential condition for the usefulness of names as descriptors, namely the check whether names are correct. Botanical names, regulated by the ICBN, can be checked by comparing with the protologue. If an accession has been found to be correctly named, a symbol can be added behind the name, as is done in the seed indexes of botanic gardens.

Cultivar names, regulated by the ICNCP, cannot be verified by studying protologues. However, the check whether cultivar names have been correctly applied is made possible by living and conserved standards, descriptions and memory.

A _living standard_ has to be sampled from the original

plant material. With regard to old cultivars, a sample has to be taken from the accession at the time of introduction. With regard to modern cultivars, a living standard has to be sampled preferably from the original plant material, grown at the winner's nursery. This is already done in several cases by registration authorities.

The same holds for conserved standards. They should consist of herbarium specimens, colour photographs, colour slides and preparations.

By description, the oldest, validly published description is meant, irrespective of the acceptance of the cultivar by a registration authority.

The memory of experts is sometimes the only way to identify an accession as a named cultivar.

The variation of cross-pollinated plants necessitates living and conserved standards, containing enough variation to circumscribe a certain cultivar and to distinguish it from other similar cultivars. In comparison with living standards, the rate and nature of drift and contamination can be determined after regeneration. Both phenomena may cause the shift of frequencies of characters in such a way, that an accession no longer resembles the original cultivar. In the long run accessions can only be compared with conserved standards, because living standards have a restricted size and lifetime. Consequently, determination of drift and contamination will then be impeded.

Apart from comparing with standards, accessions can be checked on original descriptions. In some cases, however, descriptions, especially of old cultivars, cannot be used because they are either non-existent or incomplete. Then it will be necessary to base identification on the expert's experience. It is obvious that this is less desirable since it is the most subjective identification.

The results of these four procedures can be used as descriptor states. Incorporation can be done by adding a four digit code to the descriptor 'cultivar name' (Figure 3). A positive check on each of these procedures will give the corresponding digit; a negative check the digit '0'.

During each regeneration of an accession every precaution should be taken to preserve its integrity. A sample of the regenerated accession has to be checked again with the four procedures on characters with descriptive value. If, for instance, the accession no longer corresponds to the living standard, the code 0-2-3-4 has to be added to the cultivar name, as can be seen from Figure 4. Changes in the four digit code must be added as soon as possible to the descriptor 'cultivar name'.

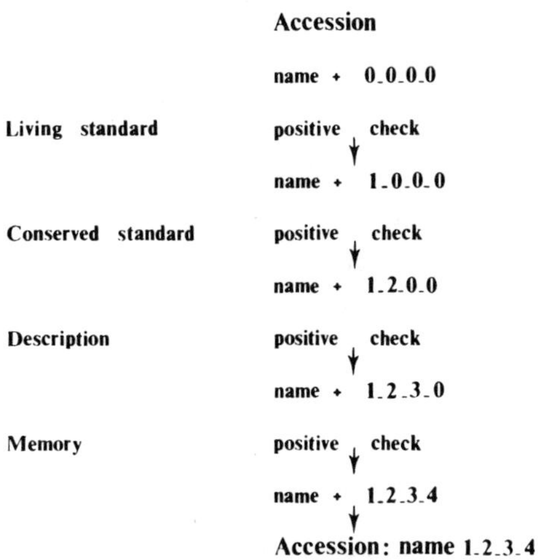

Fig. 3. Procedures to check the identification of accessions.

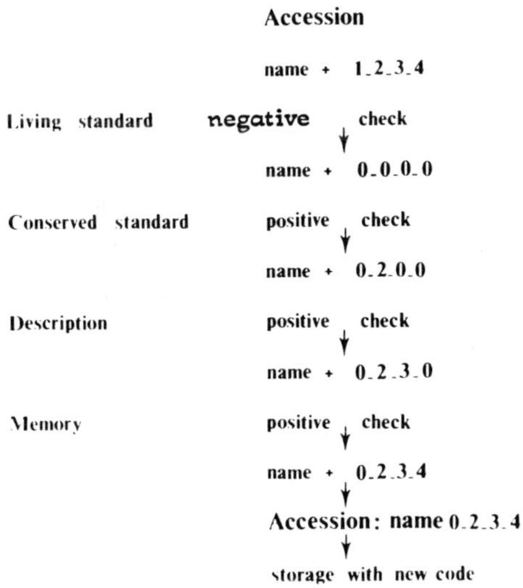

Fig. 4. Procedures to check regenerated accessions. In this example the accession will be recoded.

DISCUSSION

According to Hyland (1970), the controlled use of wild or cultivated plant germplasm is largely dependent on accurate description and taxonomic identification. Taxonomically well-identified accessions will obviously provide us with important data. Exchange of accessions will be facilitated by good, standardised documentation (Seidewitz, 1973, 1979; Erskine and Williams, 1980). As duplicates can be avoided, and regeneration especially of cross-pollinated plants is very laborious, a restricted number of well-documented accessions will be favourable, avoiding the necessity of pooling more or less similar accessions into 'race reservoirs'. A 'race reservoir' will end up as an amorphous mass without the opportunity to maintain interesting, special properties of single accessions (Marshall and Brown, 1975; Burton, 1979). In the case, however, of several poorly documented similar accessions, based only on memory (code 0-0-0-4), pooling them will be justified.

The proposed method will provide every user of GRC accessions with information about the present status of the name. A well connected name will supply entries in all documentation of the accession concerned, and will facilitate international communication.

ACKNOWLEDGEMENTS

The authors wish to express their appreciation to Dr. A.C. Zeven for his helpful remarks and to Mr. J.S. de Block, who kindly corrected the English manuscript.

REFERENCES

Ayad, G., Toll, J., and Esquinas-Alcazar, J.T., 1980. Directory of germplasm collections III Cereals 2. Maize. International Board for Plant Genetic Resources, Rome.

Burton, G.W., 1979. Handling cross-pollinated germplasm efficiently. Crop Science, 19, 685-690.

Erskine, W. and Williams, J.T., 1980. The principles, problems and responsibilities of the preliminary evaluation of genetic resources samples of seed-propagated crops. PGR Newsl., 41, 19-33.

Harlan, J.R., 1975. Crops and man. American Society of Agronomy. Crop Science Society of America, Madison, Wisconsin. p. 138, 163.

Hyland, H.L., 1970. Description and evaluation of wild and primitive introduced plants. In: O.H. Frankel and J.H. Bennett (eds.). Genetic Resources in Plants, 413-419. IBP Handbook no. 11. Blackwell Scientific Publications, Oxford and Edinburgh.

International Code of Botanical Nomenclature, 1978, adopted by the Twelfth International Botanical Congress, Leningrad, July 1975. F.A. Stafleu c.s. (ed.). Regnum Vegetabile vol. 97. Bohn, Scheltema and Holkema, Utrecht.

International Code of Nomenclature for Cultivated Plants, 1980, formulated and adopted by the International Commission for the Nomenclature of Cultivated Plants of the I.U.B.S., C.D. Brickell c.s. (ed.). Regnum Vegetabile vol. 104. Bohn, Scheltema and Holkema, Utrecht. Dr. W. Junk BV, Publ., The Hague.

Marshall, D.R. and Brown, A.H.D., 1975. Optimum sampling strategies in genetic conservation. In: O.H. Frankel and J.G. Hawkes (eds.). Crop genetic resources for today and tomorrow, 123-127. Cambridge University Press, Cambridge, London, New York and Melbourne.

Seidewitz, L., 1973. Thesaurus for the international standardisation of genebank documentation. Institut für Pflanzenbau and Saatgutforschung der Forschungsanstalt für Landwirtschaft, Braunschweig-Völkenrode.

Seidewitz, L., 1979. Need for a standardised information system. PGR Newsl., 39, 11-12.

DISCUSSION

Henny Roelefsen *(Netherlands)* I would like to know more about the four digit code you mentioned. Is it meant for computer administration?

W.A. Brandenburg *(Netherlands)* Yes, it is possible to incorporate it in a computerised system but I am not a computer specialist. To have such a code adds the descriptor cultivar name so you have the name of the accession, and the present status of the name will be given in the code. For instance, for a good cultivar which is clearly documented, you can have the code 1-2-3-4, there is a living standard, there is a conserved standard, there is a good description and there are enough people who know the accession or the cultivar. Then you have a complete code. However, if after three or four regenerations the living standard is lost because you cannot regenerate the seeds of these living standards, then you have to say the present status of your cultivar name is 0-2-3-4. This will prevent names going on and on in cases where the material is not the same as the original. That is the meaning of this code.

Henny Roelofsen So there is no reference to the documentation material in it, the code only establishes whether or not it exists?

W.A. Brandenburg Yes, it is an administrative code.

Session 2
Technical aspects: Forage plants

Chairmen: J.JAKUBIEC & J.T.WILLIAMS

Regeneration of germplasm collections of forage grasses and legumes

E.L.BREESE & B.F.TYLER
Welsh Plant Breeding Station, Aberystwyth, UK

INTRODUCTION

The domestication of crops is essentially a history of plant collection and regeneration combined with unconscious or conscious selection. Unlike the major grain crops, the history of the successful domestication of most forage grasses and clovers is telescoped into the last fifty years. The consequences of this, and the reasons for it, have profound implications for the strategies and practice of germplasm conservation.

The consequences are that many existing varieties are but little differentiated from the ecological populations ('land races') from which they arose. By the same token, continuing collections occasionally turn up germplasm which is superior to that in current use, and which can be released as commercial cultivars with a minimum of further selection and breeding. For the more advanced livestock systems, however, this is becoming an increasing rarity. Breeders are now more concerned with specific adaptations which can be incorporated into existing varieties by appropriate breeding methods as a means of extending the adaptability of the crop. Although changing agricultural patterns are beginning rapidly to erode the natural resources, there is still currently a unique wealth of available germplasm in these crops. This sets special problems as to what we should try to conserve as well as how we should seek to achieve it.

The reasons why the domestication of forage crops is so recent are particularly important in the context of this meeting. No doubt they can in part be related to the sufficiency of natural pastures for semi-nomadic practices or for lower stock intensities. Furthermore, these pastures, by virtue of their genetic heterogeneity, are highly flexible, and have the capacity to undergo some degree of evolutionary adaptation to cater for more recent higher fertiliser inputs and increased stocking densities. This is the basis of the ecotypic differentiation which allowed Stapledon (1933) and his colleagues to develop the first grass and clover varieties. However, attempts to regenerate through seed superior grasses from pastures with a high reputation for animal production date back some hundreds of years (Beddows, 1953). The attempts foundered because continued regeneration by seed resulted in a rapid loss of those vegetative characters which rendered the plants valuable as feed for ruminant animals. Only with advances in genetic knowledge early this century came the realisation that this genetic instability was related:

(i) to the outbreeding system of the species and the high genetic variation which could thus be contained within populations, and

(ii) to the conflicting physiological strategies of sexual and asexual reproduction (seed production versus vegetative growth) in these perennial species, which generated strong and opposing selection forces at the different reproductive phases.

These two aspects, reproductive mode and mating system, together with the chromosal organisation comprise the genetic system of the species which, in turn, determines the genetic structure of the various populations. However, the genetic system itself is under genetic control and may be modified by selection so that different populations of the same species may vary in genetic system and structure (Mather, 1973; Breese, 1973). As pointed out by Frankel and Hawkes (1975), a knowledge of populations structure is of fundamental importance in determining conservation strategies, both in terms of sampling during collection and maintenance. Nor can these two aspects be isolated, although in the context of this symposium we are clearly more concerned with regeneration and maintenance through seed. In considering the problems, possible solutions and ultimately the compromises enforced by practical considerations, we shall draw heavily on our experience with wind pollinated grasses, particularly the ryegrasses (*Lolium* spp.), and to a lesser extent with the insect pollinated forage legumes.

POPULATION STRUCTURE AND SAMPLING TECHNIQUES

Patterns of adaptive variation and collection strategies

Because of the importance of grassland in world agriculture, grassland species have been the subject of wide ecological studies. Many species have a widespread distribution and study reveals a vast array of adaptations to various combinations of climatic, edaphic and biotic factors. Adaptation to a particular ecological niche rarely involves response to a single environmental factor or incurs a single physiological or morphological change; rather it depends on the co-ordination of very many physiological processes in securing growth and the full reproductive cycle, with due regard to a whole complex of environmental factors. Differences of significance to the plant populations therefore tend to be quantitative rather than qualitative, depending on the fine adjustment of many genes or gene sets. Thus allelic differences of individual genes tend to be less important than specific gene combinations. This is an important point to which we shall return.

Differentiation within the species will match the pattern of environmental change. With gradual environmental changes such as those due to climatic parameters, the differentiation will be clinal (Bradshaw, 1972). This is borne out by the studies of Cooper (1964) who finds parallel adaptation in a number of species to varying light and temperature conditions from north to south latitudes. Superimposed on these broad trends will be responses to sharp discontinuities in the environment. These will be provided by biotic factors, e.g. differential grazing patterns (Breese and Charles, 1962) or by abrupt changes in edaphic factors, e.g. heavy metal tolerances (Jain and Bradshaw, 1966) or other soil factors (Tyler and Chorlton, 1976).

In extending genetic resources at the Welsh Plant Breeding Station both these have been taken into account:

(i) by mounting collection expeditions to climatically contrasting geographical areas, and

(ii) identifying discrete econiches within these areas and sampling each separately to provide representative populations (accessions).

The genetic control of variability within and between populations

A paramount consideration both for collection strategies and in sampling techniques is the way heritable variation is contained and organised within populations as well as distributed between populations (Oka, 1975; Marshall and Brown, 1975).

Many forage species have efficient self incompatibility systems which enforce a high degree of cross-pollinations and populations of these are thus expected to contain large amounts of potential genetic variability in heterozygotic and balanced homozygotic states as well as in the free state (i.e. are genetically heterozygous and heterogeneous) (Mather, 1973). Just how great this variability can be for quantitative characters controlled by many genes is well illustrated by the classical selection experiments of Cooper (1959) with perennial ryegrass. He took two contrasting natural or semi-natural populations:

(a) Irish Commercial which had been harvested by seed over many generations and was characterised by being early flowering, with a high proportion of flowering stems and relatively non-persistent,

(b) Kent Indigenous which was harvested directly from old, heavily grazed pastures and was characterised as being late flowering, leafy and highly persistent under grazing.

Selection was then practised in both populations for early and late heading (inflorescence emergence). Response was rapid in both populations and divergence was considerable before directional selection resulted in losses of fertility which acted as a brake to further response (see also Mather and Harrison, 1949 with Drosophila). A correlation between the date of heading and the degree of tillering meant that reproductive modes (sexual versus asexual) were affected simultaneously. As a consequence predominantly 'sexual', relatively short-lived, lines were obtained from Kent ryegrass and predominantly 'asexual', more persistent lines were obtained from Irish - and vice versa. The response was however more rapid and more extreme in the Kent ryegrass, a phenomenon that we will consider later.

Most importantly these experiments illustrate the vast amount of potential genetic variation that exists within single populations of these out-breeding grasses; indeed, within very small samples of plants since Cooper obtained good response from as low as four basic individuals per population. Over the years evidence has accumulated to demonstrate the scope for selection within populations for a wide range of physiological characters - for factors affecting nutritive quality, water stress and cold tolerance and for disease (for reviews see Cooper and Breese, 1971; Wilson, 1981; Breese, 1981). In no instance is there good evidence that the characters useful in agriculture or in ecological adaptation are qualitative, controlled by one or a few genes, even in the case of disease resistance where this has been researched (Wright and Breese, 1966; Hayward 1977). Rather, they are quantitative, controlled by many genes of greater or smaller effect. Consequently selection for one character may engender correlated responses in other characters because of complex linkage relationships as well as through pleiotropic effects of the same genes.

What does this imply for the strategy of plant collection and conservation? In the first instance, we can see that the main difference between populations which are ecotypically differentiated is the maximising of those plus alleles of the polygenic system which predispose adaptation to the chief limiting factor of the different environments. At the same time evolution will have enforced multi-trait selection to ensure that other characters are recombined and harmoniously balanced in linkage equilibrium. This largely dispenses with the idea that we can view populations from the point of view of allelic diversity at loci governing qualitative characters (Marshall and Brown, 1975) - rather that the important differences are in terms of different assemblies of many genes, governing quantitative characters.

As a corollary to this it is clear that these considerations completely put out of court the proposal that genetic variability can be usefully preserved as 'mass reservoirs' formed by the bulking of collections in composite crosses (Simmonds, 1962). This would rapidly lead to the break up and loss of valuable gene combinations. Particularly, in these perennial crops, it could disrupt the delicate balances between asexual and sexual reproduction, leading to a disintegration of the diversified genetic system and population structures which are presently evidenced in such genera as *Lolium* (Breese, 1973). Such practice would, in effect, dismantle the genetic edifice erected during adaptation and evolution. This might be necessary for plant breeding but is unacceptable in the preservation of genetic resources.

Number of sites and numbers of plants per site

From the above, it follows that to secure the maximum amount of genetic variability for conservation we need to collect as many contrasting habitats as possible. Limits on the number of sites should be compensated for by a greater contrast between habitats, remembering that for these quantitative characters new genetic variability may be reconstituted in crosses between populations.

The minimum number of plants that should be sampled to form the basis of each population is critical. For practical reasons which will emerge later, we have taken 25 as the minimum number. Applying the formulae of Oka (1975) this allows approximately a 93% probability of obtaining genes with a frequency of 10% in the population. Bearing in mind that the genes for the important adaptational characteristics of the populations will have been maximised, this should be adequate. By the same token, if these populations are to be maintained by inter-crossing, this number should avoid limitations on random mating imposed by too few incompatibility alleles and at the same time prevent too close inbreeding.

REGENERATION OF COLLECTIONS: GENERAL PRINCIPLES

We must now turn to the practicalities of how these populations can be regenerated for long term conservation, and increased for evaluation and distribution, while maintaining genetic integrity.

Since we are concerned with plants that achieve perenniality through asexual vegetative reproduction we have the opportunity to maintain basic plants deriving from collections in the vegetative state. Indeed this has long been the practice for the basic plants of synthetic varieties; but the practice has served to emphasise the problems. There are new possibilities in this area afforded by advances in tissue culture techniques and we shall consider these briefly later. At present, however, the method of greatest practicality is through seed both as a means of storage and for increase.

The method of seed increase offered is limited by the mating systems of the species. Most temperate grasses and clovers have efficient incompatibility systems (the grasses controlled by two loci and the legumes by one) which prohibits the use of inbred lines and largely enforces the use of interpollination.

A number of possibilities exist, including systems of controlled pair or chain crossing as discussed elsewhere in this symposium by Demarly and by Roelofsen. These latter systems may require only limited facilities but they tend to be labour-demanding. Here we consider regeneration by random interpollination among a representative sample of basic individuals from each of the populations (using principles formulated for the maintenance and multiplication of synthetic varieties). In order that the integrity of the populations is preserved we have the following requirements:

1. Avoid contamination by foreign pollen through isolation.
2. Avoid random drift by using a sufficient number of basic plants.
3. Secure random mating.
4. Avoid shift through natural selection, particularly for superior seed production.

5. Ensure maximum yields of high quality seed so as to reduce the number of subsequent regenerations and hence minimise risks from 1 - 4 above.

Isolation

For wind pollinated crops, Griffiths (1950) has shown that the extent of pollen dispersal decreases sharply with distance so that after 200 m the degree of contamination downwind is of the order of 5%. For large seed stands this contamination may be further reduced by discarding the outer metre band, since pollen mass within the stand effectively dilutes residual foreign pollen. This latter contingency is not possible for small seed islands. However, tall growing cereal crops may effectively act as pollen screens and isolation islands may be effectively established at intervals within such crops. Alternatively we may consider custom built, pollen proof, isolation glasshouses as discussed later.

For the bee-pollinated herbage legumes, studies by Roberts (1967) indicate that contamination between seed crops can be as low as 2% with distances of only 25 yards, but this is highly dependent on the positioning of the hives, and the density of flowering in the crops. Safe isolation distances with clovers tend to be recommended as of the same order as for grasses (i.e. 200 m). Again for small seed lots, insect proof glasshouses provide the greatest safeguards.

Numbers of basic plants

We have already argued that 25 basic plants provide a sufficient safeguard against serious genetic drift, while at the same time obviating problems of too few incompatibility alleles or too close inbreeding. We shall see later that this number can be accommodated in small isolation chambers and also has logistical advantages in handling.

Non-random pollination

With wind pollinated species there are always problems in securing the maximum intercrossing between representative plants of the populations, which depends greatly on the distance between plants in relation to the strength and direction of air currents. For the small number of selected basic plants of synthetic varieties, maximum intercrossing under field conditions is secured by cloning the plants and adopting an orthogonal layout to ensure that genotype contiguity is randomised (Wright, 1962). Even where numbers of genotypes in the population are large, a degree of non-randomness may arise from assortative mating (e.g. for flowering time). Evidence of this has been produced by Hayward and McAdam (1977) using isozyme markers. The seriousness of this, unless accompanied by selection, is difficult to determine in terms of gene loss from the population, or in terms of evaluation and description. This is clearly an area that requires more research.

Population shift

This is undoubtedly the most serious hazard during seed production from perennial forage crops, where survival and herbage production under grazed conditions depend on characters which are the antithesis of those characters which predispose good seed production (particularly the proportion of fertile to vegetative tillers in grasses, or stolons versus seed heads with clover). Indeed, as pointed out earlier, potentially valuable ecotypes (land races) of leafy, persistent pasture type perennial ryegrass, showed rapid shift to stemmy and short-lived types in early attempts to commercialise through seed production. In present day commercial cultivars drastic shift is circumvented by restricting the numbers of generation of seed multiplication, and by growing and harvesting the seed crops under carefully controlled conditions. The crops are carefully inspected with special reference to isolation and for prevention of mechanical admixtures during drilling and harvesting.

Even so shift can occur as evidenced by data produced by Breese and Thomas (unpublished) for four experimental late flowering ryegrass varieties over 3 or 4 generations of seed increase at varying sites in the UK (Table 1). Significantly the greatest shift in three of the varieties is towards earlier

TABLE 1

GENETIC SHIFT IN GENERATIONS OF SEED MULTIPLICATION IN PERENNIAL RYEGRASS

	Generation	Mean date of ear emergence (June)	No. of heads in aftermath growth	Water soluble* carbohydrate (%)	Germination (%)
Ba 8903	1	16.9	0.9	25.9	79.6
	2	13.2	1.7	25.1	87.0
	3	12.3	3.4	24.3	88.5
	4	10.0	4.7	23.5	90.7
Shift		-6.9***	3.7 N.S.	-2.4***	
Ba 8904	1	17.0	7.3	19.1	79.4
	2	15.3	7.9	20.5	90.1
	3	11.0	11.9	20.9	89.7
	4	7.8	16.3	21.6	93.2
Shift		-9.2***	9.0 N.S.	+2.5***	
Ba 134	6	9.7	9.5	-	64.5
	7A	7.1	7.5	-	93.5
	7B	6.3	11.2	-	96.3
	8	4.2	10.6	-	83.5
Shift		-5.5***	3.7 N.S.		
Ba 8416	1	10.9	6.9	-	93.2
	2	11.2	4.3	-	80.7
	3	11.0	3.5	-	92.9
Shift		-	-3.4 N.S.		

*** $p < 0.001$

Data: E.L. Breese and A.C. Thomas, and *I.B. Abdullah (1980)

flowering together with a greater degree of flowering head production in the aftermath (although the latter is non-significant in the data presented). Thus despite wide precautions there is a tendency for selection towards genotypes with a greater predisposition for seed-head (inflorescence) production and, as we know from previous experiments, a consequent reduction in the vegetative tillering which underlies persistency and leafiness in these late flowering varieties. The fourth population (Ba 8416) showed less change which underlines the fact that the degree of genetic instability is population-dependent even in late flowering material and may reflect past selection histories. Thus two of the populations, Ba 8903 and Ba 8904, had been previously selected over 3 or 4 generations for high and low leaf water-soluble carbohydrates respectively and response to this had been accompanied by decreased seed fertility in the case of the high line Ba 8903. It would thus be expected that relaxed selection for sugars during seed multiplication would result automatically in selection for improved fertility and a consequent reduction in the selected character in this line (see Mather and Harrison, 1949). It is noteworthy that both the high and low lines showed a regressional loss of the selected character (see Table 1), showing that genetical stability during seed multiplication is not simply a function of seed fertility and production although these may be major causes.

The third population, Ba 134, in fact represents stored seed from multiplications of a pasture ecotype (land race). This too shows shift. It is worthwhile noting that the stored seed (from generation 6) showed a declining germination percentage (Table 1), and Griffiths and Pegler (1964) have demonstrated that an increasing proportion of viable plants from such seed lots may show losses in fertility. This has considerable implications for the long term storage and regeneration of seed, which needs further researching.

The above results illustrate the particular problems of regeneration by seed in the late flowering, predominantly asexual ryegrasses from heavily grazed situations. This problem is probably less severe in the earlier flowering hay-type perennial ryegrasses and particularly the shorter lived Italian ryegrasses (as discussed by Breese, 1973). Similar considerations apply to other forage species which are differentially adapted to survival predominantly by seed or by asexual vegetative reproduction. This again illustrates the importance of a knowledge of population structure in these considerations.

The above raises many problems as to how we may best minimise shift in the small base populations of plant collections. A number of features are salient. First, the areas and climate of seed production should be chosen so as to:

(i) avoid selective elimination by virtue of plant kill by winter cold or summer drought,

(ii) have suitable weather for harvesting under optimum conditions.

Second, the timing and method of harvesting is critical, bearing in mind that most wild populations will demonstrate differences between plants in maturity dates. Third, the density of the plants in the seed stands can be important (E.W. Bean, personal communication), since the greater the density the greater the potential for competitive elimination, or at least the competitive suppression of seed heads of the relatively 'shy seeders'.

These three aspects again advocate the use of glass isolation chambers - where there is a degree of environmental control, with the plants grown at low density so as to encourage maximum seed production from individual genotypes, and where the maximum amount of ripe seed can be harvested from each individual with a degree of independence of dates of maturity.

We shall now consider these aspects in relation to seed regeneration of grass collections as practised at Aberystwyth.

AIMS AND PRACTICE IN SEED REGENERATION OF GRASS COLLECTIONS AT THE WELSH PLANT BREEDING STATION

Practical aspects in this section are considered in greater detail by Tyler in a later paper in this conference.

The original accessions of forage plants may be in the form of seed systematically collected from a number of seed heads distributed over the identified econiche; or they may be in the form of vegetative material (tillers) again collected systematically so as to be representative of the population. Where possible, the latter method has a number of obvious advantages in perennial forage crops. Thus populations can be sampled:

(i) when seed formation in the natural state is prevented through grazing,

(ii) independent of the seasonal timing of seed production, following from which

(iii) selection for specific maturity dates is avoided, and

(iv) selection for a greater predisposition to flowering is avoided.

Additionally contamination by pollen from adjacent stands is avoided. The method does however provide additional problems in handling the material and in quarantine arrangements

In both cases, seed or vegetative material is collected from not less than 25 genotypes per site and usually more (up to 30 genotypes in the case of vegetative material, up to 100 genotypes for seed).

In either case the aim will be to produce first generation seed of high quality which is sufficient to provide:

(i) basic seed for long term storage,

(ii) stock seed for initial spaced plant and small plot evaluation, and for supplying small samples to breeders and researchers.

These considerations are paramount in determining the tactics of regeneration, particularly in maximising the amount of seed which can be produced from isolation entities of minimum size. As a practical target we aim for 100 g of seed per collected ryegrass population; of which 10 g is designated for long term storage under low temperature (-20°C) conditions, 30 g is used for initial evaluation, and 60 g is allocated for medium term storage (0°C ± 2°C) for supplying demands by breeders and researchers. Seed production targets vary with the other species.

Growing the plants

Some 30 - 50 plants of each collection are grown from individual seed or separate (genotype) tillers in 18 cm pots in glasshouses. Supplementary light is provided to facilitate growth. Where seed rather than vegetative material is collected, special conditions may be necessary for germination which may only be determined by experiment; e.g. certain cocksfoot populations from northern Spain require alternating temperatures in order to induce full germination (Bean, 1976). The collection of vegetative material on the other hand will impose special quarantine restrictions which may require their being grown over a period of several months, in insect proof glasshouses under appropriate supervision.

During the initial growing period, different populations of different species may have specific vernalisation requirements of low temperatures and/or short days in order to induce fertile reproductive heads (Cooper, 1960). If the former, it is important to secure that the temperatures do not fall so low as to bring about plant kill. Again there may be specific day length requirements for full floral development (Cooper, 1950).

Isolation of the populations

For the ryegrasses, induction requirements will have been met over winter either outside on standing platforms or in cold glasshouses. Isolation requirements are provided by a series of compartments in a glasshouse, which are kept at positive air pressure by compressor fans, thus excluding external foreign pollen. The pressure is maintained by blowing air through a series of ducts to each chamber, the air itself passing through a series of pollen proof sieves. Air change within the chamber is at the rate of 60 an hour and serves to keep temperatures at acceptable levels on hot days. Air currents generated by the pressurised supply through the ducts are also designed to ensure pollen dispersion throughout the chamber and so aid in obtaining random mating. This dispersion is, however, not always as effective as might be hoped and it may be necessary to supplement the dispersion by agitation of the flowering heads. Details of the chambers may be supplied on request.

Ninety-six of a smaller type of isolation chamber are available for the multiplication of smaller seed lots including original plant collections. The chambers have a floor area of 1.4 m x 1.2 m and this is sufficient to accommodate at least 25 plants at spacings of 18 cm. On average, plants grown at this density can give a total seed yield for the population of more than 100 g for ryegrass, although this can vary from 25 g to 120 g.

Harvesting the seed

To facilitate through-put of populations, individual plants can be enclosed in pollen proof bags after pollination is virtually complete, and removed outside for the final ripening. This allows later flowering populations to occupy places in the pollination chambers. The seed is harvested by hand and bulked on threshing. Competition may exacerbate differences between plants in seed production, although growing the plants at low densities minimise this effect, quite large differences still remain in the seed productivity of the different genotypes. Bulking the seed thus alters the maternal genotypic balance in the subsequent generation which would be additional to differences in paternal genetic contributions through unequal pollen production. There is a strong argument, therefore, for mixing equal contributions of seed from each maternal genotype, at least for the basic seed lots which are designated for long term storage: thereby minimising genotypic selection and shift for longer term conservation strategies. At Aberystwyth we are considering the logistics and labour involved in the latter, while bulking all seed for stock intended for initial evaluation and for shorter term conservation and dissemination to breeders and researchers.

Seed quality and seed storage

For both short and long term storage there is a strong need for the highest quality seed. Experiments by Bean (1974) have shown that seed production under glass in the way described is superior in terms of quality as well as yield to other methods of artificial isolation (e.g. polythene tunnels; pollen proof bags) and is more reliable than seed production under natural conditions.

Our approach to the drying and storage of the seed follows the principles laid down by Roberts (1975). Generally speaking the seed are dried to 5 - 7% moisture, and for medium term storage are sealed in air-tight containers and lodged in cold rooms maintained at $0^{o}C \pm 2^{o}C$, or in unsealed containers in cold rooms with 20% relative humidity. For long term storage, cold cabinets held at $-20^{o}C$ are being used. It is not the purpose of this paper to consider in detail the technical problems of seed storage, but the length of time that seed can be stored without genetic deterioration and loss of germination is obviously critical to a consideration of the frequency of regeneration. There is particularly a need for research of the extent and type of genetic deterioration even with moderate reductions in germination as prognosticated by the experiments of Griffiths and Pegler (1964) quoted earlier (see also Roberts, 1975).

SCOPE FOR GENETIC CONSERVATION THROUGH ASEXUAL REPRODUCTION

Many perennial grasses and clovers can be maintained potentially indefinitely through cloning and this would circumvent the inherent genetic instability of these outbreeding organisms during sexual reproduction. However, under natural conditions it is not only a laborious and expensive operation but it is also fraught with risks of loss from virus infection, disease and other natural hazards.

Recent developments in tissue culture techniques now offer increasing scope for long term storage of vegetative material (Morel, 1975; Henshaw, 1975). Thus virus free plantlets from meristem cultures of Italian (Dale, 1980a) and hybrid ryegrasses have now been successfully produced. These can be maintained in disease free conditons at 2 - $4^{o}C$ in small tubes containing agar with basic nutrients designed just to sustain growth. The plants remain slow growing and small and need reculturing only at infrequent intervals (a year to 18 months). With a change of medium, particularly the inclusion of cytokinins, the plants can be induced to tiller rapidly and thus regenerate numerous ramets. This clearly has a high potential not only for the clean storage of genetic stocks but as a tactic in the production of large quantities of first generation seed. The basic plants of an increasing number of our varieties are now being maintained in this way. Similarly encouraging results are being obtained in meristem cultures of clovers (Cheyne and Dale, 1980).

In a number of plant species there is now the possibility of deriving callus, cell and protoplast cultures which can be regenerated into differentiated plants. Some success has been obtained in regenerating grasses from callus and perhaps single cells and protoplasts (Vasil and Vasil, 1980; Dale, 1980b). In some forage legumes successful regeneration has repeatedly been obtained from protoplasts (e.g. Kao and Michayluk, 1980). However, such cultures often give rise to great genetic variability (D'Amato, 1975; Shepard et al., 1980) which although possibly of great use to plant breeders does not provide the stability necessary for long term conservation of genetic stocks.

Even with the inherently greater stability of meristem cultures, we need to investigate more fully the potential for variation in 'mericlones', bearing in mind the potential for somatic selection demonstrated in certain clones of perennial ryegrass by Breese et al. (1965).

FREQUENCY OF REGENERATION OF COLLECTIONS

As previously mentioned, risks of shift, drift and contamination are reduced by restricting the frequency of regeneration by seed. This may be sought through ensuring maximum production of high quality seed from each population commensurate with the dictates of space, labour and technical considerations which is then stored under optimum conditions. But the seed supply will vary with the genotypic potential of each population; while the demand for the seed will depend on the immediate requirements for the characteristics which they offer. Clearly the best way to cater for longer term gene conservation as well as more immediate exploitation is to establish two seed lines in the way previously indicated:

(i) Basic seed, which will be designated for long term conservation in deep freeze cabinets. The frequency of regeneration here will be determined by fall off in viability. Such stored seed will require continued monitoring for germination percentage, and there is a need for further research into the extent that viable seed shows genetic deterioration during storage. Our present target is up to 20 year cycles.

(ii) Stock seed, which is allocated for initial evaluation and distribution to breeders and researchers. This will require less sophisticated conditions for medium term storage. One further generation of multiplication of this stock seed may be considered if low seed production or heavy demands justify this (see Figure 1). This may occur after 4 - 5 years.

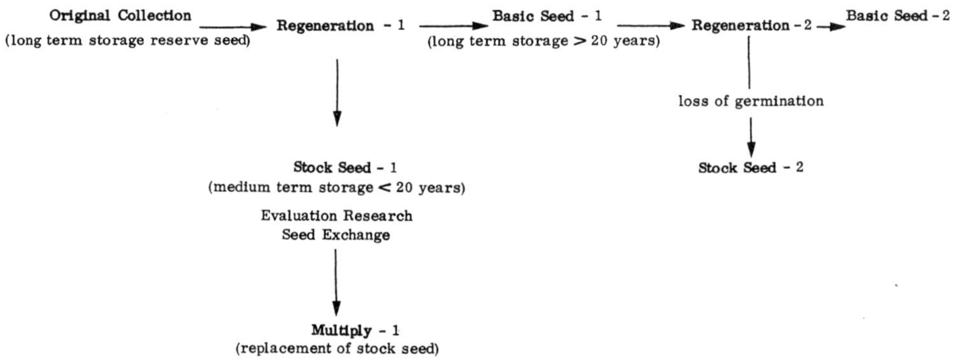

Fig. 1. Regeneration strategies.

This obviously has considerable requirements in terms of facilities and staffing. At Aberystwyth both are presently limiting. Primarily we have been maintaining what may be described as a 'breeder's collection', where the focus has been more on the immediate requirements of the breeding and research programmes. The areas from which we have collected reflect this, as do the critical characteristics we have attempted to conserve through selected populations. We have nevertheless also attempted to collect and conserve grass and clover populations from areas which are under particular threat of genetic erosion from changes in land use. This has latterly projected us more into the role of an 'active gene centre'.

CONCLUSIONS

A good knowledge of the genetic system of the species and the genetical structure of the populations provides necessary guidance in collection strategies and on regeneration by seed or other means. For grasses and legumes, experience in breeding new varieties and their multiplication for commerce has provided invaluable knowledge on the technical aspects of population regeneration and conservation through seed or by vegetative means. This experience has highlighted an urgent need for further research in almost all areas of consideration if we are to develop the most effective strategies and tactics in forage plant conservation. Not least in these considerations with regard to persistent grasses and clovers which have evolved under, and are maintained by, livestock systems, is the extent to which conservation may be obtained *in situ*, in addition to, or in complementation of, long term seed storage.

REFERENCES

Abdullah, I.B., 1980. Genetic homeostasis and population structure in *Lolium*. Ph.D. Thesis, University of Wales.

Bean, E.W., 1974. The development of quality of forage grass seeds produced under different isolation conditions. Ann. appl. Biol., 78, 179-186.

Bean, E.W., 1976. Problems of germination in ecotypes of *Dactylis glomerata*. Report of the Welsh Plant Breeding Station for 1975, p.53.

Beddows, A.R., 1953. The ryegrasses in British agriculture: a survey. Bulletin No.17, Series H, Welsh Plant Breeding Station, Aberystwyth.

Bradshaw, A.D., 1972. Some of the evolutionary consequences of being a plant. Evolutionary Biology, 5, 25-47.

Breese, E.L. and Charles, A.H., 1962. Population studies in ryegrass. Report of the Welsh Plant Breeding Station for 1961, p. 30-34.

Breese, E.L., Hayward, M.D. and Thomas, A.C., 1965. Somatic selection in perennial ryegrass. Heredity, 20, 367-379.

Breese, E.L., 1973. Genetic architecture and adaptation in grasses with special reference to *Lolium*. International Meeting on Quantitative Inheritance, Polymorphisms, Selection and Environment. University of Bologna 1972, pp. 73-81.

Breese, E.L., 1981. Exploitation of genetic resources through breeding in *Lolium* spp. In: Proceedings of the Symposium 'Genetic Resources of Forage Plants'. Ed. R.A. Bray and J.E. McIvor, CSIRO, Melbourne (In press).

Cheyne, V.A. and Dale, P.J., 1980. Shoot tip culture in forage legumes. Plant Science Letters, 19, 303-309.

Cooper, J.P., 1950. Day length and head formation in the ryegrasses. J. Br. Grassld. Soc., 5, 105-112.

Cooper, J.P., 1959. Selection and population structure in *Lolium*. III. Selection for date of ear emergence. Heredity, 13, 461-479.

Cooper, J.P., 1960. Short day and low temperature induction in *Lolium*. Ann. Bot., 24, 232-246.

Cooper, J.P., 1964. Climatic variation in forage grasses. I. Leaf development in climatic races of *Lolium* and *Dactylis*. J. appl. Ecol., 1, 45-61.

Cooper, J.P. and Breese, E.L., 1971. Plant breeding: forage grasses and legumes. In: Potential Crop Production. Ed. P.F. Wareing and J.P. Cooper, pp. 295-318, Heinemann, London.

Dale, P.J., 1980a. A method for *in vitro* storage of *Lolium multiflorum* Lam. Ann. Bot., 45, 497-502.

Dale, P.J., 1980b. Embryoids from cultured immature embryos of *Lolium multiflorum*. Z. Pflanzenphysiol., 100, 73-77.

D'Amato, F., 1975. The problem of genetic stability in plant tissue and cell cultures. In: Crop Genetic Resources for Today and Tomorrow. Ed. O.H. Frankel and J.G. Hawkes, pp. 333-348, Cambridge Univ. Press.

Frankel, O.H. and Hawkes, J.G., 1975. Genetic resources in the past ten years. In: Crop Genetic Resources for Today and Tomorrow. Ed. O.H. Frankel and J.G. Hawkes, pp. 99-110, Cambridge Univ. Press.

Griffiths, D.J., 1950. The liability of seed crops of perennial ryegrass *(Lolium perenne)* to contamination by wind borne pollen. J. Agric. Sci., Camb., 46, 19-38.

Griffiths, D.J. and Pegler, R.A., 1964. Effects of long term storage on pollen fertility and seed setting in S.23 perennial ryegrass. Rep. Welsh Pl. Breed. Stn. for 1963, pp. 81-83.

Hayward, M.D., 1977. Genetic control of resistance to crown rust *(Puccinia coronata* Corda) in *Lolium perenne* L. and its implications in breeding. Theor. Appl. Genet., 51, 49-53.

Hayward, M.D. and McAdam, N.J., 1977. Isozyme polymorphism as a measure of distinctiveness and stability in cultivars of *Lolium perenne*. Z. Pflanzenzüchtg., 79, 59-68.

Henshaw, G.G., 1975. Technical aspects of tissue culture storage for genetic conservation. In: Crop Genetic Resources for Today and Tomorrow. Ed. O.H. Frankel and J.G. Hawkes, pp. 349-358, Cambridge Univ. Press.

Jain, S.K. and Bradshaw, A.D., 1966. Evolutionary divergence among adjacent plant populations. I. The evidence and its theoretical analysis. Heredity, 21, 407-441.

Kao, K.N. and Michayluk, M.R., 1980. Plant regeneration from mesophyll protoplasts of alfalfa. Z. Pflanzenphysiol., 96, 135-141.

Marshall, D.R. and Brown, A.H.D., 1975. Optimum sampling strategies in genetic conservation. In: Crop Genetic Resources for Today and Tomorrow. Ed. O.H. Frankel and J.G. Hawkes, pp. 53-80, Cambridge Univ. Press.

Mather, K., 1973. Genetic Structure of Populations. Chapman and Hall, London.

Mather, K. and Harrison, B.J., 1949. The manifold effect of selection. Heredity, 3, 1-52, 131-162.

Morel, G., 1975. Meristem culture techniques for the long term storage of cultivated plants. In: Crop Genetic Resources for Today and Tomorrow. Ed. O.H. Frankel and J.G. Hawkes, pp. 327-332. Cambridge Univ. Press.

Oka, H.I., 1975. Consideration on the population size necessary for conservation of crop germplasms. In: JIBP Synthesis Vol. 5 - Gene Conservation. Ed. T. Matsuo, pp. 57-63. Univ. of Tokyo Press.
Roberts, E.H., 1975. Problems of long term storage of seed and pollen for genetic resources conservation. In: Crop Genetic Resources for Today and Tomorrow. Ed. O.H. Frankel and J.G. Hawkes, pp. 269-296. Cambridge Univ. Press.
Roberts, H.M., 1967. Isolation requirements in white clover seed crops. Rep. Welsh Pl. Breed. Stn. for 1966, pp. 103-105.
Shepard, J.F., Bidney, D. and Shahin, E., 1980. Potato protoplasts in crop improvement. Science, 208, 17-24.
Simmonds, N.W., 1962. Variability in crop plants, its use and conservation. Biol. Rev., 37, 442-465.
Stapledon, R.G., 1933. An account of the Organisation and Work of the Station from its Foundation in April 1919 to July 1933. Welsh Plant Breeding Station, Aberystwyth.
Tyler, B.F. and Chorlton, K.H., 1976. Ecotypic differentiation in *Lolium perenne* populations. Rep. Welsh Pl. Breed. Stn. for 1975, pp. 14-15.
Vasil, V. and Vasil, I.K., 1980. Isolation and culture of cereal protoplasts. Theor. Appl. Genet., 56, 97-99.
Wilson, D., 1981. Breeding for morphological and physiological traits. In: Plant Breeding. II. Ed. K.J. Frey, Iowa State University.
Wright, C.E., 1962. A systematic polycross design. In: The Research and Experimental Record of the Ministry of Agriculture, Northern Ireland XI. Part 1, pp. 7-8.
Wright, C.E. and Breese, E.L., 1966. The genetic control of blind-seed disease in *Lolium perenne* in relation to breeding practice. Proc. 10th Int. Grassld. Congr. Helsinki 1966, pp. 737-741.

DISCUSSION

G. Jenkins *(UK)* You mentioned the possibility of variation arising in protoplast regeneration as we know from Shepard's work in potatoes. Is there much evidence of the same phenomenon in grasses?

E.L. Breese *(UK)* We have not, as yet, gone so far as to be able to discover this. The work with calluses is only just beginning - we have quite a lot of callus but we have not looked at the kinds of instability that may derive from it. We are just getting to the stages in Italian ryegrass where we think we might be getting callus from individual cells. In the grasses we have not yet got regenerations from the protoplast, the basic cell. As you know, in the legumes this is now increasingly possible; it has been achieved in lucerne and we are hoping that we will get right down to the protoplast in the clovers as well.

I should add one thing here. Our own experience earlier on cloning, and the ability to select within clones for response, tells us that we should be very careful to look even at that relatively uncomplicated meristem tissue to see whether in the so-called 'mericlones' we have some shift as well. We are checking on this in our storage methods of the basic plants for varieties, but so far we have not noticed any great shift here.

E. Porceddu *(Italy)* You say that at your Institute you have succeeded in selecting for both high and low tillering from the same plant, and that some of the genetic control will be passed from the genome to the cytoplasm or the plasmon. When you now preserve seeds, thus forcing plants to produce seeds, would you expect to lose some of the control that was probably the result of centuries of adaptation?

E.L. Breese In the experiment where we attempted to see whether there was variation which could be exploited by selection in a clone, we took a number of different clones. Our responses tended to be largely confined to those plants which had a greater predisposition to asexuality. They were the highly tillering pasture types. We tried it also in perennial ryegrass plants that had been derived from successions of seed multiplications and our success here was not conclusive; we did have some suggestion but it was certainly much less evident. The speculation we advanced at this time was that in the very persistent strains which regenerated chiefly by asexual means, there was some delegation by the nucleus to plasmon-based mitochondria or other DNA or RNA bodies of this ilk - plasmids or whatever. One presumes that some of this characteristic may be lost, yes, if we continue to reproduce through seed, but so little is known about this that I would hesitate to predict exactly how much of it we would lose. I would hope that tissue culture work will be able to clarify this.

B.H. Somaroo *(Syria)* You have outlined a scheme for regeneration of the grasses. Is this the same scheme you follow for the legumes or is it very different?

E.L. Breese I would think that the scheme in itself is that which we would try to pursue with the legumes. As yet we have done very much more work with, and had very much more experience of, the grasses. Our colleagues in Aberystwyth, the legume breeder and the seed multiplication department, do have quite a lot of experience with clover at least. Largely speaking with the variants of the species, the experience tends to confirm the principles of what I enunciate here and I would think that we would try to mount a scheme with legumes which is largely forming the kind of structure that we have put here. Perhaps Bruce Tyler would like to comment further on that.

B.F. Tyler *(UK)* The big difference, of course, is the difference between insect and wind pollinated plants. The big practical problem is to maintain bumble bees in an active way in isolation chambers. However, I think there is a little bit of progress. It means going round with test tubes collecting bumble bees or collecting nests which is a practical problem, but work is being done on this.

E.L. Breese Yes, obviously the practicalities are different but I think there are people here, such as David Astley, who can probably comment more precisely on the problems of maintaining insect pollinated plots.

D. Astley *(UK)* For our out-pollinated vegetables, we have tended to move over to standard blowflies rather than bees. That is not so easy for legumes but for crops such as *Brassicas* and *Alliums* it is very easy. There is no selective pollination from the insect, it tends to be a random pollination and in terms of total seed production flies are just as effective as bees. There are other considerations: you do not need a beekeeper or course, flies are much easier to handle. We purchase maggots from a commercial fishermen's shop. We are lucky that the British are a nation of fishermen and so these are very cheap. We are able to pupate the maggots and then hold them in suspended animation at +1 or +2 until we actually require the flies for pollination. So, overall they are a very effective pollinator. IVT Netherlands also use this system.

E.L. Breese In summary, I think we can say that the strategies would generally be the same but the tactics clearly would be different. Our experience with legumes and bees indicates that it is practical in pollination controlled insect proof housing and that is what I think we would generally be aiming for.

Practical aspects of a regeneration scheme in forage grasses

B.F.TYLER
Welsh Plant Breeding Station, Aberystwyth, UK

INTRODUCTION

The primary aim of the 'working collection' at the Welsh Plant Breeding Station is to provide genetically diverse, characterised populations showing differences in major agronomic and physiological characters.

Although initially our activities were concerned mainly with collection and agronomic characterisation (evaluation), (Davies et al., 1973), it soon became apparent that an adequate supply of high quality seed was essential both for breeding and research, as well as for an active seed exchange service. This involved development of a scheme for regeneration, seed storage and a computerised storage/retrieval system for the associated data. The following paper considers some of the operational aspects of the scheme, which is described elsewhere in this Symposium (Breese and Tyler). Figure 1 illustrates the procedure schematically.

COLLECTION

First it is necessary to consider our collection techniques, since these influence considerably the methods used in the initial regeneration.

The objective of the collection is to obtain as representative a sample as possible within the limits set by practical constraints. Thus both seed and vegetative sampling are possible. Both methods have merits and demerits and these have been considered previously.

Seed

This method is less time consuming on site and also poses fewer problems of maintenance during the expedition. However,

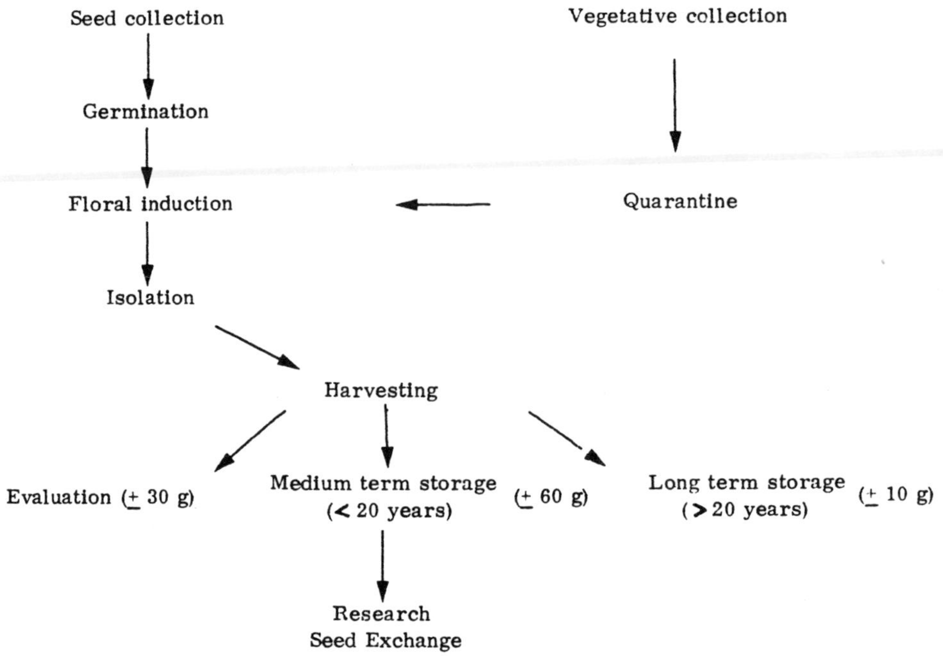

Fig. 1. Regeneration procedure.

the very real possibility of bias towards the selection of freely flowering genotypes, when the population to be sampled is grazed or trampled, makes this method less suitable when collecting the mainly vegetatively reproduced *Lolium perenne*. Seed collections are usually confined to other forage grass species particularly when management intensity is low or when the reproductive phase is allowed to develop unhindered (e.g. hedgerows, uncultivated land etc.). The aim is then to collect up to 100 heads, randomly distributed over the whole site; taking a single head at each point to ensure equal genotypic representation, and with at least 10 metres between points to avoid duplicating genotypes.

If it is suspected that the population may contain a mixture of ploidy levels (e.g. *Dactylis* collections from north-west Spain; Chorlton et al., 1978), it is important that the heads are kept separate so that individual chromosome counts can be made and the different chromosome races kept distinct. Apart from this consideration seed heads from each population are bulked.

Vegetative

This method is often preferred particularly when sampling perennial, mainly vegetatively reproduced ecotypes. It is often the only possible method if intensive grazing or trampling have prevented the production of seed heads. Thirty vegetative units are sampled in an analogous way to seed collections. These are placed in polythene bags, sealed and stored in cool, insulated containers and the temperature maintained between 16 - 20°C (Tyler and Chorlton, 1978).

For both methods we aim to maximise genetic differences between populations, while restricting variation within populations as much as possible. In practice this means identifying discrete ecological areas within sites and sampling each separately. The value of such 'micro-environmental' differences in obtaining genetically differentiated populations has been emphasised by the work of Bradshaw (1960), Tyler (1970), Tyler and Chorlton (1976).

SEED REGENERATION

Growing the plants for seed production

The aim here is to provide optimum growing conditions to avoid, as far as possible, selective suppression of individual genotypes.

Seed collection

Approximately 100 seeds are sown either in boxes of JI compost in a glasshouse or on germination pads, depending on the quality of seed and its anticipated germination. Special requirements for germination such as alternating temperature (Bean, 1976) or pre-chilling (Tyler et al., 1978) are sometimes necessary. Thirty seedlings, taken at random, are potted on in 10 cm pots.

Seed collections are invariably made in July or August and germination and subsequent growth occurs in September through to November. Under these conditions of shortening day length and low light intensity, it is necessary to use supplementary lights so that sufficient growth has occurred to enable the plants to be given conditions for floral induction. We aim to have the plants of sufficient size, hardened and placed outside under natural day length and temperature, by mid-December. Collections brought in from warmer regions may need protection against cold in order to avoid selection.

Vegetative collection

Transfer of live material from overseas to UK requires import permits and has to be maintained in suitable quarantine houses, under supervision.

On return to the home institute all plants are sprayed with a systemic insecticide and fungicide and placed in a quarantine house. The quarantine house is an insect proof, sectionalised glasshouse ventilated with filtered incoming and outgoing air.

Each vegetative unit (divot) is reduced to a single tiller and planted in a box of JI compost. This ensures that the subsequent seed island is composed of a finite number of different genotypes. Growing conditions in the quarantine house can be controlled to some extent by the provision of heat and the use of portable lights. The plants are transferred to 10 cm pots as soon as sufficient growth has occurred. Any supplementary light or heat is removed by mid-December in order to approach the short day length and low temperatures required for floral induction.

In March of the following year plants derived from either seed or vegetative units are repotted into 18 cm pots.

Isolation and harvesting

During this phase the aim is to produce a maximum yield of high quality seed from each participating genotype with the exclusion of foreign pollen. In order to approach this ideal, 25 of the original 30 plants are randomly chosen and transferred to isolation chambers immediately prior to anthesis.

The isolation chambers used are small (floor area 1.4 m x 1.2 m) adjacent units of aluminium alloy and glass. Each chamber is made pollen-proof, air-tight and weatherproof by using a PVC seal. Ventilation is by fan-blown filtered air ducted to each chamber such that a positive outward air pressure is maintained. The air outlet is through a filter in the door; air change within the chamber is at the rate of about 60 changes an hour. A second thermostatically controlled compressor fan is set to cut in at around 17°C to provide additional air movement and cooling during periods of high temperature.

Water is provided by capillary action using felt matting placed over a central channel containing a perforated alkathene pipe.

The island is arranged in a 5 x 5 block with a spacing of 18 cm between plants, thus leaving additional space for smaller islands of pair crosses of non-compatible species in the chamber.

In a breeding institute heavy demands are imposed on these isolation facilities not only for regneration of genetic resources but also for the initial seed increase of breeders lines, so that rapid throughput is essential. In order to effect this, plants are removed from isolation about 15 days after peak anthesis, the heads being covered with pollen-proof bags. It is thus possible to isolate later flowering populations of the same species in the same season.

This method of early removal of islands before the seed is completely formed could affect the seed contribution from the genotypes, particularly in that extreme early and late flowering genotypes may be disadvantaged. This would of course magnify other inherent differences in seed production among the genotypes. In order to minimise such differences in the formation of a bulk seed lot, equal seed amounts from each genotype should be used so as to maintain the maternal balance. Practically, however, this would result in the seed yield of the island being governed by the lowest yielding genotype, and often insufficient seed would be available for evaluation. In addition, individual harvesting is extremely time-demanding and operational decisions have to be made in the light of available resources. A compromise is often possible where 0.5 g is taken from each genotype and bulked to form the basic seed for long term storage and the remaining seed is bulked for evaluation and seed exchange.

Threshing, cleaning and storage

After harvest, seed is dried either in a glasshouse or in the laboratory in open mesh cotton bags. A drying room where dry air is continually circulated would be ideal. Dry seed is essential for efficient threshing. The seed heads are initially pounded to remove most of the seed and then mechanically threshed to remove the remainder. Hand-sieving and mechanical blowing provide a clean, high quality seed sample. The bulk seed is then transferred to drying cabinets where the seed is dried passively over silica gel to 5 - 7% moisture. After drying, the seed is weighed and divided up to provide approximately 10 g of basic seed for long term storage at -20°C. The rest of the seed is constituted into stock of which 30 g is used for evaluation and the remainder put into medium term storage at 0°C ± 2°C to provide for further research and seed exchange. The seed for storage is sealed in laminated foil envelopes using a crimp heat-sealer.

SUBSEQUENT REGENERATION

The preceding scheme aims initially to produce high quality seed of sufficient quantity for conservation and immediate use such that the need for regeneration is reduced to a minimum. However, the requirements for breeding, research and seed exchange of popular ecotypes may require further seed regeneration. Regeneration for this purpose would be from the stock seed, stored under medium term conditions.

Basic seed stored at -20°C is anticipated to maintain viability for long periods (over 20 years) before rejuvenation is necessary. Substantial loss of viability resulting in possible genetic change of the remaining seed by differential gene loss and by increase in mutation rate (Roberts, 1975) indicates that continual monitoring of each accession for viability is essential. This is a responsibility of major genebanks and not possible with the limited resources available in 'working collections'. However, random monitoring of basic stocks is possible and a regeneration cycle is implemented when indications of a loss in viability are found.

CHARACTERISATION AND DATA STORAGE

The value of a collection, in our experience, depends ultimately on the amount and nature of characterisation that has been possible on each accession.

Data on accessions are derived from two main sources: i) information obtained during collections; ii) information obtained during characterisation and evaluation. The former is likely with all collections, the latter only either from a working collection or as data fed back to a base genebank from its users. In both cases an efficient computerised storage system, ideally with interactive retrieval, is necessary.

CONCLUSIONS

Regeneration from plant collections is necessary to provide i) stock seed for initial evaluation and for supplying breeders and research workers and ii) basic seed for long-term conservation. The latter seeks to preserve existing genetic stocks for future generations of breeders, whose requirements may be for characteristics quite different from those presently studied. It is thus essential to maintain the integrity of each population as far as possible. In seeking this during seed regeneration, inevitably, there has to be a compromise between the theoretically desirable and that which is practically possible. At the WPBS we have developed an operational scheme based on research results presently available and which is commensurate with the limitations of staff and facilities. In this sense the scheme itself is experimental. The methods must continually be reviewed in the light of new research results (as discussed earlier in this symposium by Breese and Tyler) and the technique modified according to the facilities offered.

REFERENCES

Bean, E.W., 1976. Problems of germination in ecotypes of *Dactylis glomerata*. Rep. Welsh Pl. Breed. Stn for 1975, p. 53.

Bradshaw, A.D., 1960. Population differentiation in *Agrostis tenuis* Sibth. III. Populations in varied environments. New Phytol. 59, 92-103.

Chorlton, K.H., Young, N.R. and Tyler, B.F., 1978. Extending genetic resources - collecting *Dactylis glomerata* in Spain. Rep. Welsh Pl. Breed. Stn for 1977, pp. 42-43.

Davies, W.E., Tyler, B.F., Borrill, M., Cooper, J.P., Thomas, H. and Breese, E.L., 1973. Plant Introduction at the Welsh Plant Breeding Station. Rep. Welsh Pl. Breed. Stn for 1972, pp. 143-162.

Roberts, E.H., 1975. Problems of long-term storage of seed and pollen for genetic resources conservation. In: Crop Genetic Resources for Today and Tomorrow. (Ed. O.H. Frankel and J.G. Hawkes), pp. 269-296, Cambridge Univ. Press.

Tyler, B.F., 1970. Sources of genetic variability: plant introduction. Rep. Welsh Pl. Breed. Stn for 1919-1969, pp. 16-18.

Tyler, B.F. and Chorlton, K.H., 1976. Ecotypic differentiation in *Lolium perenne* populations. Rep.Welsh Pl. Breed. Stn for 1975, pp. 14-15.

Tyler, B.F. and Chorlton, K.H., 1978. A *Lolium* and *Festuca* plant collecting expedition to northern Italy. Rivista di Agronomia 12, 181-190.

Tyler, B.F., Borrill, M. and Chorlton, K.H., 1978. Studies in *Festuca* 10. Observations on germination and seedling cold tolerance in diploid *Festuca pratensis* and tetraploid *F. pratensis* var. *apennina* in relation to their altitudinal distribution. J. appl. Biol. 15, 219-226.

DISCUSSION

Henny Roelofsen *(Netherlands)* I see that you bag the plants after the main pollination. Does that not affect your seed quality because the ripening inside the bag may be different from outside the bag?

B.F. Tyler *(UK)* We do it fifteen days after peak anthesis and there is a possibility that the quality of the seed might be affected but we have looked at it quite carefully and there does not seem to be any great fall-off in quality, although there may be a small amount. You can use transparent bags if you wish. There is always a compromise in these situations; we do have very good seed quality from these isolation houses probably because the growing conditions, the pollination conditions and the initial development of the seed are so good in relation to outside that the small fall-off caused by bagging is compensated for by the other advantages.

Henny Roelofsen I wasn't thinking about light as much as moisture content.

B.F. Tyler They are not cut, they are still on the plant.

Henny Roelofsen Yes, I realise that but the moisture content may be too high.

B.F. Tyler The bags are porous, moisture does not form inside, they are ordinary miller paper bags. There is no build-up of moisture at all. They don't have to be completely airtight because the chances of getting pollen in from the bottom are fairly remote.

Henny Roelofsen That was my second point. I wonder if it is necessary to bag them after the main flowering is over because if you leave them to dry and to ripen then there will be hardly any normal sized seeds from the latest flowers.

B.F. Tyler Well, there are problems. There is about a 30 day spread in the grasses in terms of when particular tillers develop and emerge. It is true that we do lose a certain amount of yield in this way, I admit this. The problem with not bagging them afterwards is that at a breeding institute there is so much pollen about, particularly with the ryegrasses, that it would be rather dangerous. We are quite happy with the quality of the seed, in comparison with the seed we get from the field anyway.

V. Connolly *(Ireland)* You mentioned that you take just about 25 genotypes as a minimum. Bearing in mind that the major problem is getting to the site, do you think that more than 25 might perhaps be desirable?

B.F. Tyler It depends. Again, it is a compromise between the number of people you have got collecting, the time available, and so on. When we are collecting for Hayward, when he is looking at his gene frequencies and various things like this, we do aim to get 100 plants, but that is a specialist situation. It does take up a lot of time. In fact, I think we would prefer to collect from more sites than from more plants per site. We are aiming to maximise differences between populations and, therefore, we want to have more distinct populations. Of course, if we could notice considerable environmental differences within the meadow, we would collect vegetatively within these smaller areas as well. But it is a question of compromise.

W. Plarre *(Federal Republic of Germany)* I think you have to consider the statistical point of view and I would say 25 to 30 was satisfactory from that point of view.

S. Andersen *(Denmark)* When you are multiplying your grasses in greenhouses you have them in a temperature comparable to the Mediterranean region and in a light intensity comparable to Iceland, so this could give you a selection which is not normal under British conditions. When you take an equal amount of seed from each plant it presupposes that out in the field each genotype would occupy the same space, otherwise you would make a selection there too. Therefore, my question is, have you any evidence that this complicated system is more satisfactory than sowing out in the field and collecting seed the next year?

B.F. Tyler We feel that we would have a chance of considerable selection if we sowed out in the field, through repressing seed production on some of the shy-seeding genotypes and also possible plant kill. We are aiming to get representative genotypes from the meadows in our first multiplications. Perhaps Dr. Breese would like to comment.

E.L. Breese *(UK)* Although going into the greenhouses gives a very high temperature, what we are seeking to do is to minimise selection under that temperature by low plant densities and other means, ultimately by taking equal quantities of seed from each genotype. Unless you have very specific gametic or zygotic elimination at those temperatures, you are thereby more or less guaranteeing an equal maternal effect, although not necessarily a paternal one, in the production of pollen. In that context you could have an effect at the paternal level but it is the best we can do and I am quite sure that it is less than you would get out in the field under natural conditions, with less control of the population.

Maintenance and multiplication of cultivars of allogamous field crops in Denmark

M.H.POULSEN
University of Agriculture, Copenhagen, Denmark

C.P.PEDERSEN
Daehnfeldt Ltd., Odense, Denmark

INTRODUCTION

Maintenance of breeder seed and multiplication of this seed for commercialisation of a new cultivar are important activities. They must fulfil demands as to stability, purity and quality of the seed. Before the introduction of the OECD Certification Scheme and the Plant Breeders' Right legislation, maintenance and multiplication could be, and often was, undertaken by any firm who could claim 'genuity' of their original stock of seed. Cultivars were often altered or at least adjusted to the locality where the multiplication took place. Consequently stocks of a named cultivar were not always genetically identical.

Under the current rules, multiplication of cultivars is controlled by the authorities concerned with cultivar testing and the certification of commercial seed in order to secure the purity and quality of the seed.

Maintenance of the cultivar is the obligation of the breeder and it must be carried out in a way which will secure the identity of the cultivar as originally described. Genetic drift and migration of genes from other populations must be avoided. Genetic drift is avoided by including a sufficiently high number of plants in the process. In practice the minimum number of plants often varies due to the circumstances under which the maintenance has to take place. Here the work involved with the different species is often considered in connection with the risk of genetic drift.

The multiplication is organised and controlled by the seed firms. Farmers with long experience in seed production are selected for multiplication of pre-basic and basic seed.

In the following we outline the genetic structure, the

maintenance and the multiplication of the allogamous crop species as practised in Denmark.

CULTIVAR STRUCTURE

In principle cultivars can be referred to in two classes depending on the way in which they are constructed:

1. Composite cultivars
2. Single population cultivars.

The composite or synthetic cultivars are built up of components. The components are typically offspring families selected from poly-, top- or pair crosses. Where recurrent cloning is simple, selected clones after a similar crossing and testing system may form the basis of a variety. In such cases the offspring seed from a controlled intercrossing of the clones are maintained.

Single population cultivars may be created from mass selection in another cultivar or from various segregating populations i.e. families from pair crosses etc. However, several Danish cultivars of this type are based on a limited number (10 - 30) of single plants selected from various populations and intercrossed.

STORAGE AND MAINTENANCE

Seeds of each component (family or single clone offspring) are stored in a deep freezer or similar equipment in order to keep them viable for a longer period of time. When required, the components are grown for maintenance as shown in Figure 1. In the case of a composite cultivar, 100 - 200 single plants are raised from each component. The plants from a component are grown together and isolated from other material of the same or nearly related species. Thus, for example, *Lolium perenne* and *Lolium multiflorum* are always kept apart to avoid hybridisation. The plants are controlled for earliness in heading and off types are discarded. The components are harvested separately. All or part of the seed may be stored again or it may be mixed and sown in a lay for intercrossing and production of breeder seed.

The single population cultivars are maintained in a sim-

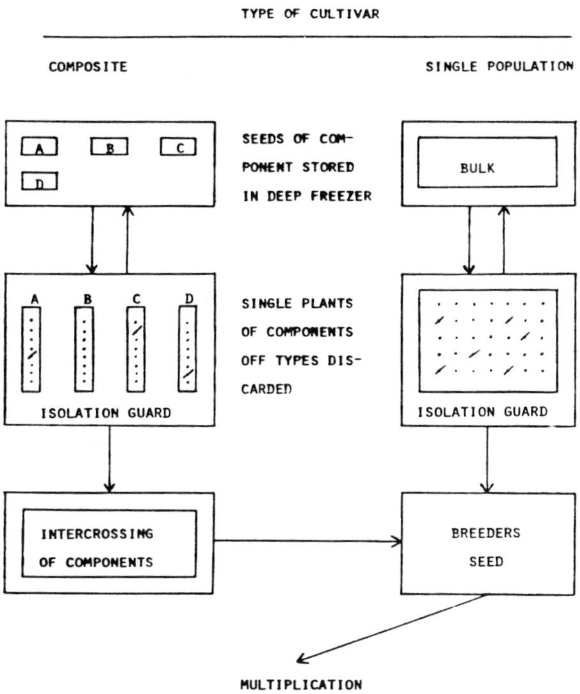

Fig. 1. Storage and maintenance of allogamous cultivars.

ilar way. However, in this case breeder seed is produced either directly from the original stock as harvested on the selected and intercrossed single plants or recurrently from breeder seed. About 1 000 single plants are raised. Due to the wider genetic bases more variation is often found in this type of cultivar compared to the variation in the composite cultivars.

MULTIPLICATION

Of breeder seed approximately 0.5 ha is laid down to produce pre-basic seed. Also pre-basic seed may be stored for a longer period in a special seed storage compartment where the temperature and the humidity of the air are kept low: 2 °C and 20% RH. Pre-basic seed is used mainly for production of basic seed from which in turn certified seed is produced. The multiplication pathways are shown in Figure 2. In the figure it is also demonstrated that regeneration of pre-basic seed as well as of basic seed, usually takes place. The frequency of such regeneration is determined according to the type of cultivar.

For composite cultivars this type of regeneration should be limited so that the certified seed is always of approximately the same generation after intercrossing the components.

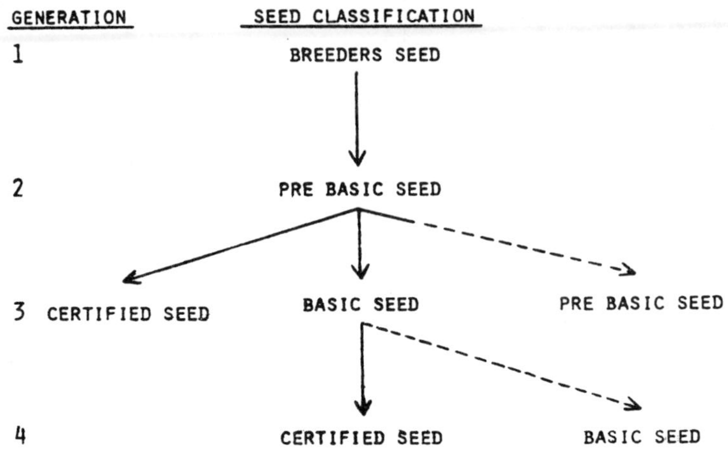

Fig. 2. Multiplication (solid lines) and regeneration (broken lines) for cultivars of allogamous species.

ISOLATION

It is obvious that contamination of the small maintenance plots as well as of the multiplication fields is unnecessary. Distance must be kept between contamination sources both in time and space.

Distance in time means that seed production must not take place on land where remnant plants or seed from a previous crop may cause a contamination source for multiplication.

In the maintenance of cultivars this problem is less important as the plots are small and therefore may be effectively controlled and cleaned by hand and/or by special equipment such as a row rotary cultivator.

For multiplication purposes a field location is chosen so that distance in time is obtained to other cultivars of the same species. A distance in time is also kept for species with similar seed size and shape to the species in question. This is especially important in the small seeded legumes, as hard seeds may be viable in the soil for several years. For example red clover may be a serious contaminant in lucerne because the seeds of these two species cannot be separated mechanically.

The certification demands in respect of distances in various crops are outlined in Table 1.

TABLE 1

DEMANDS FOR DISTANCE TO CONTAMINATING SOURCES IN MULTIPLICATION OF BASIC SEED AND CERTIFIED SEED OF SOME ALLOGAMOUS CROPS (Landbrugsministeriets certifikatudvalgs regler at 6 februar 1975).

	Distance in metres		Distance in years	
	Field 2 ha and less	Field more than 2 ha	Same species	Different species
Forage legumes				
Basic seed	200	200	7 - 10	5
Certified seed	100	50	3	3
Grasses				
Basic seed	200	200	5	5
Certified seed	100	50	2	3
Beets				
Basic seed	1 000		8	4
Certified seed	800		8	

Introduction and use of the modern grass herbicides (Round Up etc.) has more or less solved the problems associated with surviving grass plants. It is now possible to clean the land for surviving grass 'weeds'. Although this has reduced the demand for a distance in time when multiplying grasses, proper crop management is required.

Distance in space is necessary to reduce gamete migration with pollen. Experiments with collection of air borne pollen have shown that grass and beet pollen are spread over considerable distances. As shown in Table 2, Jensen and Bøgh (1941) found that the amount of pollen decreased to 10% over a distance of 500 m. However, at a distance of 1 000 m 5% was still found. The figures are percentages of the amount found at the edge of the field, consequently in areas where intensive seed production takes place the amount of foreign pollen in the air is considerable and migration of foreign genes is inevitable. The certification scheme with its demands on field location and implemented control, ensures that high purity is obtained.

TABLE 2

PERCENTAGE OF POLLEN GRAINS FOUND AT AND WITH INCREASING DISTANCE (WITH THE WIND) FROM A FIELD OF WIND POLLINATED SPECIES (From Jensen and Bøgh, 1941).

Distance from field	Species *Lolium perenne*	*Phleum pratense*	*Dactylis glomerata*	*Beta vulgaris*
0	100	100	100	100
100	19	41	35	27
200	14	25	27	16
300	14	16	18	14
400	7	11	13	9
500	4	8	5	6
600	-	-	6	5
800	-	-	6	6
1 000	5	-	3	-
1 200	5	-	-	-

The general impression is that bees are less effective in spreading the pollen than wind. However, experiments with lucerne show that considerable contamination (10 - 30%) can be expected even when the distance to other sources of the same species is 150 m (Johansen, 1968). In field bean Pope and Bond (1975) under English conditions, found very small contamination (1.25% and 0.59%) at distances of 92 m and 194 m between small plots.

Use of the solitary leaf cutter bees *(Megachile rotundata)* may be an advantage in the multiplication of small fields because of the limited foraging radius of these bees.

The maintenance plots of grasses, forage legumes and beets are placed at specially selected locations. Here woodlands, hedges and buildings are considered to provide a better isolation than just distance in metres. In addition to the careful choice of location an additional windbreak is often provided by growing winter rye around the plots of grasses and clovers. In the case of regeneration of the single components of a composite cultivar of grasses it is usual to plant the single plants into plots situated in a winter rye field. The small plots (50 m^2) are situated approximately 10 m apart and are fully surrounded by the rye.

REFERENCES

Jensen, I. and Bøgh, H., 1941. Om Forhold, der har Indflydelse paa Krydsningsfaren hos vindbestøvede Kulturplanter. Tidskr. f. Planteavl 46, 238-268.

Johansen, B.R., 1968. Isolation Distances in Lucerne. - Royal Veterinary and Agricultural University, Yearbook, Copenhagen 1968, pp. 153-158.

Pope, M. and Bond, D.A., 1975. Influence of isolation distance on genetic contamination of field beans *(Vicia faba* L.*)* seed produced in small plots. - J. agric. Sci. Camb. 85, 509-513.

DISCUSSION

E.L. Breese *(UK)* In relation to isolation distances, you took this on pollen traps I believe. Dr. Griffiths' experiments were finally done with markers, using the recessive lack of anthocyanin in the base. As you point out, there was a tremendous difference according to whether one used small collections of widely spaced individual plants or larger areas of dense stands. With individual plants with no break in between, or small groups of about six plants, then you have the kind of graph that I showed with contamination reducing to about 5% after which it became constant at about 200 m. However, if you had a dense stand, particularly if you removed, say, the first 3 m of the margin of the field nearest to the contaminant source, then you could reduce the area to about 50 yards. In other words, the intra-varietal density of pollen had the same effect as distance and could in some cases be equated with it. As I understand it the recommendations are now with us: about 200 m for breeder seed, 150 - 200 m for basic seed, but down to about 100 m for certified seed. It is of that order but I am not sure of the exact figures.

Incidentally, with bees, I think that Roberts discovered that depending on the hive source and the density of clover heads, contamination could be as little as 2% over 25 yards from the source. Bees have quite a local form of distribution.

M.H. Poulsen *(Denmark)* I would just like to make one comment about bees. Clearly we agree that bees are less effective in spreading pollen than the wind but there are data from Denmark showing that under certain circumstances the situation can be different. The example I am thinking of was a big field of lucerne with a dominant gene. At increasing distances from that, a recessive type of lucerne was sown in small plots. On these small plots there was tremendous contamination from the big field. Of course, it would not be sensible to have the multiplication or maintenance of small plots in close proximity to a large field, so these figures may not be entirely relevant, but there was about 30% contamination.

Pollination and seed regeneration in lucerne

F.VERONESI
National Research Council, Perugia, Italy

F.LORENZETTI
University of Perugia, Italy

E.PORCEDDU
National Research Council, Bari, Italy

Lucerne (*Medicago sativa* L.) is a prevalently allogamous perennial species. It is native to the temperate regions of western Asia: Media, the countries of the southeast Caucasian range and the northwestern part of Iran (Whyte, 1958). Klinkowski (1933) states that it is the oldest known forage plant and it is apparently not possible at the present day to indicate a more definite area for its origin. At present, lucerne is the most important leguminous forage plant in several countries of Europe, Asia, North Africa, North and South America, and therefore there is a great need for collection and maintenance of its germplasm resources.

Before discussing the most appropriate methods for germplasm conservation, it is useful to give some information on the pollination system and genetic structure of natural lucerne populations.

The percentage of natural crossing in lucerne varies with temperature, moisture and bee activity as well as with the genotypes involved (Wilsie, 1957). In Italy, the most important pollinators seem to be honey bees (*Apis mellifica ligustica*) which visit lucerne flowers to collect nectar. Some workers believe that honey bees cause a large amount of selfing (Lesins, 1961; Hanson et al., 1964); on the other hand, other workers maintain that honey bees are important for tripping and cross-pollination (Pedersen, 1962). Honey bees seem to increase seed production per plant (Figure 1) and this seed production seems to be mainly due to cross-pollination (Table 1) in the climatic conditions of central Italy (Lorenzetti and Porceddu, 1968).

As far as the percentage of selfing and the genetic structure of natural populations are concerned, data obtained

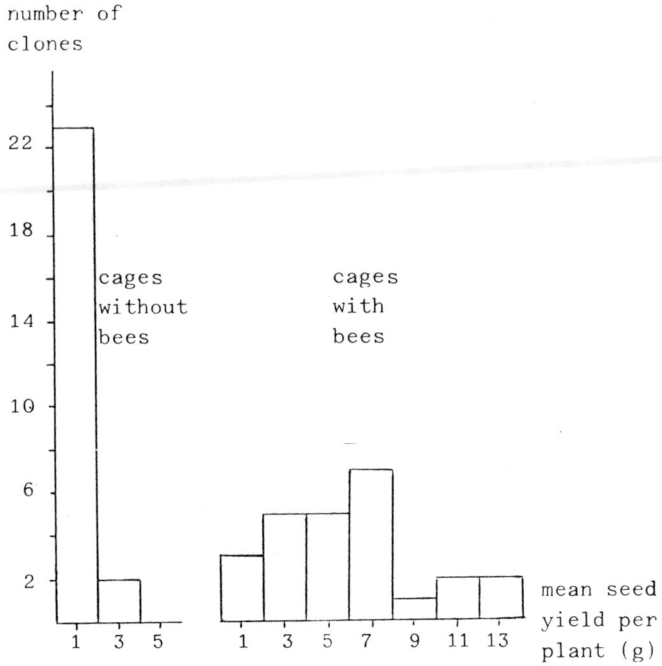

Fig. 1. Mean seed yield per plant observed in 25 clones of lucerne from cages with or without bees.
(Modified from: Lorenzetti and Porceddu, 1968).

TABLE 1

PERCENTAGES OF PODS WITH MORE THAN TWO COILS OBTAINED BY DIFFERENT TYPES OF POLLINATION FROM 5 CLONES OF LUCERNE

Type of pollination	Pods with more than two coils (%)
Selfing	14.2
Cage pollination	
without bees	15.8
with bees	37.4
Hand crossing	37.1

(Modified from: Lorenzetti and Porceddu, 1968).

by several authors, by the usual method of controlling open pollination progeny of plants homozygous for recessive markers, show an average 15% of selfing (Lesins, 1961; Mayer, 1962). For the tetrasomic lucerne pattern of inheritance (Stanford, 1951; Demarly, 1963) it is unlikely that plants homozygous for recessive markers have been originated by intercrossing rather than by self-fertilisation. On the other hand, self-fertilisation brings about a reduction in self-fertility (Tysdal et al., 1942; Brink and Cooper, 1963) so this method should lead to an underestimate of the amount of selfing.

An alternative way could be the study of the distribution curves of quantitative characters in selfed and crossed experimental populations compared with the curves of natural ones (Lorenzetti and Brandi, 1978).

The drastic drop in self-fertility in just one generation of selfing suggest that selfs growing in the field are not likely to reproduce unless they are outcrossed (Wilsie, 1957). Moreover, several authors have pointed out a marked inbreeding effect on vigour. The inbreeding depression of green matter yield is about 30% in spaced plant trials (Figure 2) and can reach 50 - 60% in solid seeding trials (Figure 3) after a single generation of selfing. In these conditions, the distribution curves of progeny obtained by selfing (S_1) or crossing (F_1) are clearly distinct while an intermediate pattern is shown by natural populations (P) (Figure 4).

Assuming that the distributions of F_1 and S_1 are normal and the natural population (P) is a mixture of S_1 and F_1, the following system of equations can be written for a quantitative trait:

$$K_{F_1} + K_{S_1} = 1$$

$$K_{F_1} \cdot \bar{M}_{F_1} + K_{S_1} \cdot \bar{M}_{S_1} = 0$$

where $\bar{M}_{F_1} = \bar{X}_{F_1} - \bar{X}_P$ and $\bar{M}_{S_1} = \bar{X}_{S_1} - \bar{X}_P$. In the system, K_{S_1} and K_{F_1} indicate the amount of selfing and crossing, respectively.[1] Experiments made in Perugia (Table 2) with populations

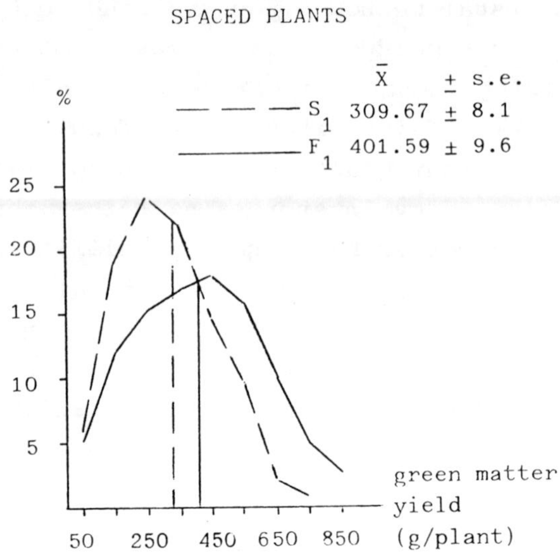

Fig. 2. Frequency polygons relative to green matter yield (g/plant) of S_1 and F_1 plants at the first cut of spring 1981 (Perugia).

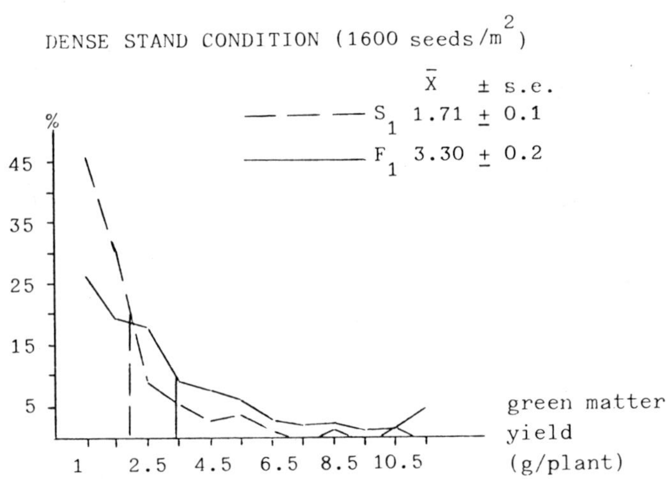

Fig. 3. Frequency polygons relative to green matter yield (g/plant) of S_1 and F_1 plants at the first cut of spring 1981 (Perugia).

obtained from local types of lucerne by:

a) controlled self-pollination (S_1);

b) controlled cross-pollination (F_1);

c) open pollination (P)

indicate that the amount of selfing in central Italy is about 30 - 40% (Lorenzetti and Veronesi, in press). Probably, the amount of self-pollination in lucerne differs considerably with the populations and with the environmental conditions. Lucerne germplasm conservation could be made by tissue culture storage, vegetative propagation or seed storage with periodical seed regeneration to avoid germinability losses and genetic degeneration.

In practice, a drawback to storage in culture is genetic variability due to mixploidy in the original explant, and to events occurring during culture initiation and maintenance. Euploid and aneuploid abnormalities may be found in plant tissue cultures and plants regenerate from culture. The abnormalities may relate to heterogeneity in the parent plant but are more likely to have developed during culture initiation and maintenance. Also if there is evidence that cultures may be stabilised by careful choice of inoculum and care in maintenance, a serious threat to genetic stability is the effect, in the long term, of background radiation. Because the subject is relatively new, further work is needed to characterise specimens before and after storage to settle questions in this area (Withers, 1980).

The vegetative propagation of lucerne involves considerable risks both for degeneration caused by virus and for populational shifts to genotypes better adapted to vegetative propagation. At any rate, in order to keep a natural population balance it is necessary to restore gamic reproduction also, from time to time, in vegetatively maintained germplasm introductions.

As far as germplasm stocks maintenance with seed regeneration is concerned, the method based on selfing is not possible because of a marked inbreeding depression effect and a large variability in self-fertility which leads to selective processes and undesirable changes in gene frequencies. Moreover selfing

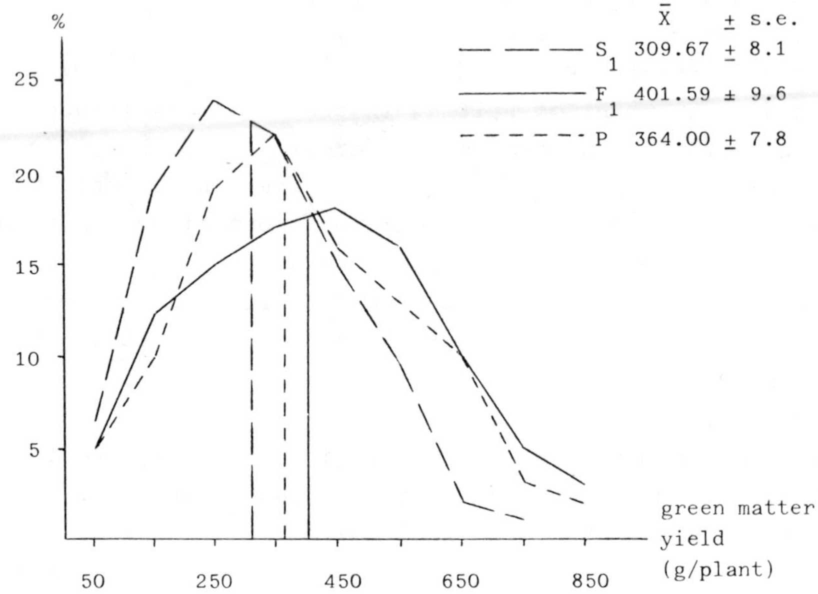

Fig. 4. Frequency polygons relative to green matter yield (g/plant) of S_1, F_1 and P (natural population) at the first cut of spring 1981 (Perugia).

TABLE 2

K_{F_1} AND K_{S_1} VALUES DETERMINED ON THE BASIS OF DATA CONCERNING HEIGHT AND GREEN MATTER YIELD PER PLANT AT THE FIRST CUT OF SPRING 1979.

Coefficients	Height	Green matter yield
K_{F_1}	0.64	0.66
K_{S_1}	0.36	0.34

(From: Lorenzetti and Veronesi, in press).

is very time consuming because of the necessity of isolating plants and self-pollinating flowers by hand.

At present, open-pollination obtained in artificial isolation with the use of bee-proof wire netting and small honey beehives seems to be the most suitable system for lucerne seed regeneration even if the problem of bee preference for some plants persists. Such a seed regeneration could be applied both in dense stand and in spaced plants conditions.

The method of regenerating a population by seed is very important to maintain the genetic variability of original materials and involves several problems in relation to the environmental conditions in which seed regeneration is made. To provide data about the action of natural selection on seed regeneration of germplasm introductions is not simple; however, a very rich bibliography exists on the behaviour of selected varieties of allogamous forage plants multiplied out of their adaptational regions (Laude and Standford, 1960; Bula, 1966; Simon, 1966; Valle, 1966; Dovrat and Waldman, 1966; Kelly and Boyd, 1966). The effects of natural selection on modifying lucerne populations were clearly explained by Smith (1955, 1958) and Zaleski (1962).

As a rule, an adapted population phenotypically selected consists of many different genotypes which are often phenotypically similar in the original environment. When this population is grown under new environmental conditions, different genotypes show a variability unknown in the region of adaptation (Ludwig et al., 1953; Clausens and Hiesey, 1958; Cooper, 1959). Seed multiplication of a population out of its original region is exposed to natural selection which determines a sort of new ecotypisation. Such a type of ecotypisation produces changes in gene frequencies due to differences in seed production from genotype to genotype and increases the allele frequencies of the most adapted genotypes after a few generations.

In addition, in allogamous perennial forages there is a selective pressure due to intraspecific competition and the decrease in plant density with time. In lucerne, as a result of competition, diseases, insects, winter injuries and other causes, the number of seedlings surviving the first year after seeding is generally in the range of 40 to 50% of the seed sown (Demarly et al., 1964; Tesar and Jackobs, 1972). Frequently, seedling survival is as low as 20% (Vivenza, 1914; Haussman, 1955; Panella, 1960; Brown et al., 1960) and this percentage further decreases in subsequent years (Figure 5). It may be believed that elimination of seedlings is only fortuitous during the first period after the seeding, but it becomes selective soon after. So, environmental selective pressures should act on seed production less in the year of seeding than later.

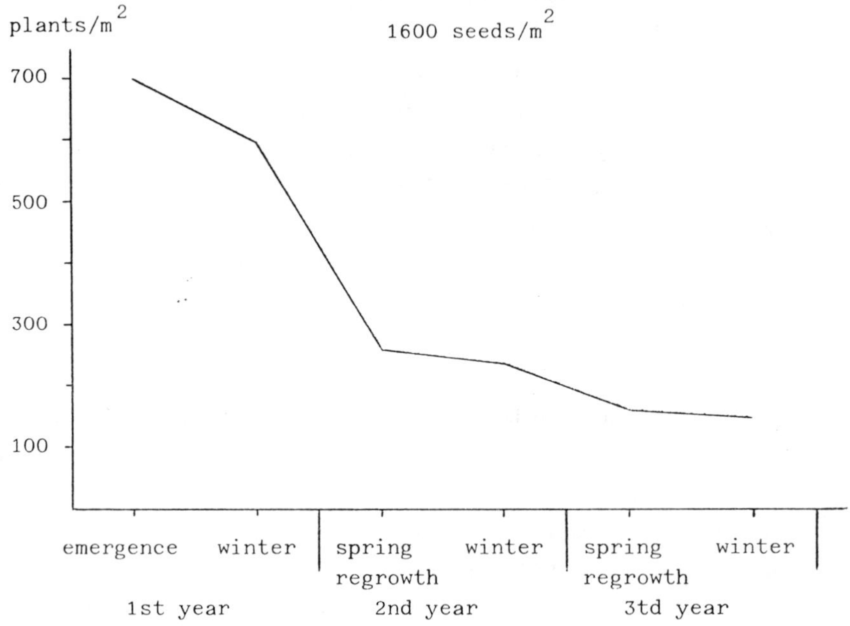

Fig. 5. Plant density in lucerne stand (Perugia: 1978-1980).

Natural selective pressure can be attenuated with the decrease of intraspecific competition. For example, with germination of lucerne seeds in Petri dishes and subsequent pricking out of seedlings as spaced plants, more than 90% of well established plants able to produce open pollinated seed are obtained. So, the absence of competition partially decreases the ecotypisation risks. In lucerne, open pollinated seed production of 300 - 400 genotypes for each population, grown as spaced plants, seems to be the most useful seed regeneration system. It should provide a long term conservation of genebanks stocks. At any rate, lucerne seed regeneration must be made in climatic conditions as near as possible to the original ones, preferably in the region of adaptation.

REFERENCES

Brink, R.A. and Cooper, D.C., 1963. The mechanism of pollination in alfalfa *(Medicago sativa)*. Amer. J. Bot., 23, 678-683.

Brown, B.A., Decker, A.M., Sprague, M.A., MacDonald, H.A. and Teel, M.R., 1960. Maryland Agr. Exp. Sta. Northeast Regional Publ. No. 41 Bull. A-108.

Bula, R.J., 1966. Plant-type comparisons of red clover, *Trifolium pratense* L., from seed produced at diverse latitude. Proc. X Int. Grassl. Congr., 783-786.

Clausens, J. and Hiesey, W.M., 1958. Experimental studies on the nature of species. IV. Genetic structure of ecological races. Carnegie Inst. Wash. Publ., 615.

Cooper, J.P., 1959. Selection and population structure in *Lolium*. II. Genetic control of ear emergence. Heredity, 13, 445-449.

Demarly, Y., 1963. Génétique des tetraploides et amélioration des plantes. Ann. Amél. des Plantes, 13, 307-400.

Demarly, Y., Guy, P. and Chesneaux, M.T., 1964. Analysis préliminaires de la compétition chez les luzernes. Ann. Amél. des Plantes, 14, 131-155.

Dovrat, A. and Waldman, M., 1966. Changes in flowering responses and seed yields in some northern forage varieties multiplied in a northern latitude. Proc. X Int. Grassl. Congr., 787-792.

Hanson, C.H., Graumann, H.O., Elling, L.J., Dudley, J.W., Carnahan, H.L., Kehr, W.R., Davis, R.L., Frosheiser, F.I. and Hovin, A.W., 1964. Performance of two clone crosses in alfalfa and an unanticipated self-pollination problem. Agr. Res. Serv. USDA Tech. Bull., 1300.

Haussman, G., 1955. Densità di semina dei medicai in rapporto al substrato. Ann. Acc. Agr. Torino, 97, 1-18.

Kelly, A.F. and Boyd, M.M., 1966. The stability of cultivars of grasses and clovers when grown for seed in differing environments. Proc. X Int. Grassl. Congr., 180-184.

Klinkowski, N., 1933. Lucerne. Its ecological position and distribution in the world. Imp. Bur. Plant. Genetics: Herbage Plants, Bull. No. 12.

Laude, H.M. and Stanford, E.H., 1960. Environmentally induced changes in gene frequency in a synthetic forage variety grown outside the region of adaptation. Proc. VIII Int. Grassl. Congr., 180-184.

Lesins, K., 1961. Mode of fertilisation in relation to breeding methods in alfalfa. Z. Pflanzenzüchtung, 45, 31-54.

Lorenzetti, F. and Porceddu, E., 1968. Honey bees and the production of polycrossed lucerne seed. Z. Pflanzenzüchtung, 60, 156-164.

Lorenzetti, F. and Brandi, P., 1978. Autoimpollinazione e impollinazione incrociata in erba medica. Annali della Facoltà di Agraria dell' Università degli Studi di Perugia, XXXII-XXXIII: 209-228.

Lorenzetti, F. and Veronesi, F. New data on the amount of selfing in lucerne. Report of Eucarpia Fodder Crops Section Meeting, *Medicago sativa* Group, Kompolt, Hungary, July 1-4 (in press).

Ludwig, R.A., Barrales, H.L. and Steppler, H., 1953. Studies on the effect of light on the growth and development of red clover. Can. J. Agr. Sci., 33, 274-287.

Mayer, R., 1962. L'amélioration des plantes en France. Ann. Amél. des Plantes, Numéro Hors Série, 63.

Panella, A., 1960. Il diradamento dei medicai e il problema della resistenza. Progresso Agricolo, VI, no. 7.

Pedersen, M.W., 1962. An analysis of certain factors associated with pollination and seed production in alfalfa. Proc. 1st Int. Symp. Pollination, 64-74.

Simon, V., 1966. Advances in growing seed of German alfalfa varieties in California. Proc. X Int. Grassl. Congr., 773-776.

Smith, D., 1955. Influence of area of seed production on the performance of ranger alfalfa. Agron. J., 47, 201-205.

Smith, D., 1958. Performance of Narragansett and Vernal alfalfa from seed produced at different latitudes. Agron. J., 50, 226-229.

Stanford, E.H., 1951. Tetrasomic inheritance in alfalfa. Agron. J., 43, 222-225.

Tesar, M.B. and Jackobs, J.A., 1972. Establishing the stand. Alfalfa Science and Technology. American Society of Agronomy, Inc., Publisher Madison, Wisconsin, USA, 415-435.

Tysdal, H.M., Kiesslbach, T.A. and Westover, H.L., 1942. Alfalfa breeding. Nebr. Agr. Expt. Sta. Res. Bull. 124.

Valle, O., 1960. Experience with seed production of Northern European forage crops grown at various latitudes in North America. Proc. VIII Int. Grassl. Congr., 185-188.

Vivenza, A., 1914. La vita di un medicaio. La St. Sp. Agr. It., XLVII, 2, 97-118.

Whyte, R.O., 1958. Plant exploration, collection and introduction. FAO Agricultural Studies. No. 41.

Wilsie, C.P., 1957. Effect of inbreeding on fertility and vigor of alfalfa. Agron. J., 49, 182-185.

Withers, L.A., 1980. Tissue culture storage for genetic conservation. IBPGR Technical Report.

Zaleski, A., 1962. Changes in the characters of lucerne varieties when grown for seed under conditions differing from those of their country of origin. J. Brit. Grassl. Soc., 17, 7-16.

Seed multiplication of cross-pollinated forage crops at ICARDA

B.H.SOMAROO & S.CECCARELLI
International Center for Agricultural Research in Dry Areas, Aleppo, Syria

INTRODUCTION

The Forage Improvement Program is one of the research programmes of the International Center for Agricultural Research in the Dry Areas established in 1977 at Aleppo (Syria). The other research programmes are the Cereal Improvement Program, the Food Legumes Improvement Program and the Farming System Program.

The mandate of the Forage Improvement Program is to increase animal production in a region characterised by increasingly substantial imports of meat, milk, cheese and butter, and by very limited forage production. Animal production is limited and subject to great fluctuations because it is mainly based on grazing semi-arid rangelands, unproductive pastures, crop residues and weeds. The increased demand for animal products due to an increasing population, has led to an increase of animal numbers with the wellknown consequences of overgrazing, land abuse and soil erosion. Therefore, the increase in animal numbers has been accompanied by a constant decline in feed availability.

In addition, the increased pressure on the land due to population growth, has led to the expansion of cereals, such as barley, into marginal lands, hill slopes, foothill areas and even to the border of the steppe. The results of this expansion of cereal crops have been been a slight increase in cereal production and severe erosion problems.

As forage production is very limited and the region covered by the ICARDA mandate is very extensive (from Morocco to Pakistan) the acquisition of a working collection of germplasm received very high priority in the early stages of development of the Forage Improvement Program.

THE FORAGE GERMPLASM COLLECTION

A list of the genera in the forage germplasm collection is presented in Table 1; the development of the collection from 1976 to 1980 is depicted in Figure 1. The germplasm accessions have been collected in several countries of the region or acquired from research organisations, germplasm centres and other sources.

Approximately one-fourth, or about 3 800 accessions, are cross-pollinated species. For many of these accessions, especially those collected in the region, only few seeds are available.

SEED REGENERATION

The multiplication of germplasm and, more generally, seed production is of considerable importance in one or more of the following steps:

a) to obtain the minimum amount of seed needed for a preliminary evaluation in adaptation trials;
b) multiplication of the more promising accessions for replicated preliminary and advanced yield trials and for regional testing and on-farm trials;
c) production of seed for long term storage of those accessions which do not have immediate value.

Seed regeneration for preliminary evaluation

Very often the seeds of some accessions are available in extremely limited quantities, and unless the seeds are multiplied, they cannot be evaluated. For example, 25 seeds of one accession of *Reseda lutea* , a species which gave interesting results in rangeland reseeding studies in Iran, have been received recently with an indication of the germination requirements. However, three seeds planted under the specified conditions failed to germinate.

When germination is not a problem, the first seed regeneration can take place successfully in a glasshouse or even in a plastic house. If temperature and/or humidity can be controlled and the flowering biology is known, artificial lighting can be used effectively to induce flowering and ensure appropriate conditions for pollination and seed-setting.

TABLE 1

FORAGE GERMPLASM COLLECTION (1980) AT THE INTERNATIONAL CENTER FOR AGRICULTURAL RESEARCH IN THE DRY AREAS (ICARDA).

Genera	Total no. of accessions	Number of accessions acquired (1980)	Number of accessions collected (1980)
Alfalfa	799	35	3
Astragalus	286	14	39
Atriplex	11	-	-
Coronilla	48	-	16
Hedysarum	8	-	-
Hippocrepis	36	-	9
Hymenocarpus	45	-	1
Lathyrus	384	11	26
Lotus	30	7	-
Lupinus	30	-	-
Medicago	2 710	95	473
Melilotus	21	-	10
Onobrychis	727	30	27
Pisum	2 793	2 483	5
Poterium	11	-	3
Scorpiurus	45	-	12
Trifolium	796	10	44
Trigonella	136	-	4
Vicia	2 693	56	102
Barley	973	-	-
Triticale	1 333	156	-
Grasses	420	80	12
Total	14 356	2 977	786

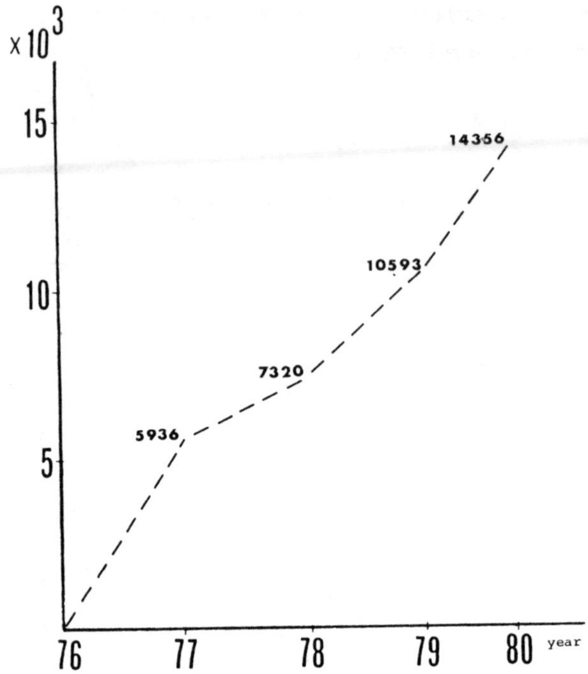

Fig. 1. Development of ICARDA's Forage Germplasm Collection

With cross-pollinated species, different accessions of the same species cannot be multiplied in the same glasshouse unless some kind of isolation is available. This can be in the form of compartments, in which accessions of different species are multiplied together, or in the form of pollination bags.

When pollination bags are used, a certain amount of assortative mating is unavoidable and it is uncertain to what extent the population obtained by artificial crossing will be affected by genetic changes. Although disassortative mating is technically feasible, and is the method which best holds populations together (Allard, 1960), it becomes impracticable when one is dealing with a large number of populations.

Another limitation in the use of pollination bags lies in seed quality. If seed quality is affected, and seedling characters are of importance, the use of seed produced under pollination bags may lead to unrealistic and incorrect evaluations.

A particular case for seed regeneration in cross-pollinated species is when the amount of seed initially available allows a preliminary evaluation for adaptation. After the more promising accessions have been identified, the same material used for evaluation can be used to increase the seed for further testing. The alternative is to use the remnant seed, if still available, to plant isolation plots or to transplant the material in isolation plots both of which are time-consuming procedures and require additional land and manpower resources.

As an example of this kind of situation some data obtained in a preliminary evaluation trial with 80 accessions of 12 species of perennial grasses will be presented. The experiment was sown at the ICARDA site at Tel Hadya (near Aleppo, Syria) in December 1979; due to the high number of varieties and/or populations and to the low amount of seed available (often as low as 40 - 50 seeds) an augmented design (Federer, 1956) which included a check population every five plots was adopted. Observations were made on summer dormancy and winter regrowth characteristics.

The results of these observations are reported here for *Dactylis glomerata* which was the species with the highest number of populations (Table 2).

TABLE 2

SUMMER DORMANCY (0 = DORMANT - 5 = GREEN) ON 14/16.4.1980, AND WINTER REGROWTH (0 = MIN - 5 = MAX) ON 12.1.1981 AND 28.1.1981 OF 27 POPULATIONS OF *DACTYLIS GLOMERATA* OF DIFFERENT ORIGINS EVALUATED AT TEL HADYA (NEAR ALEPPO, SYRIA.

Origin	No. of populations	Summer dormancy	Winter regrowth 12.1.81	28.1.81
Foreign varieties	9	2.6	1.2	1.4
Syria	12	1.1	3.0	4.1
Jordan	4	0.5	3.8	4.3
Turkey	2	2.5	1.0	3.0
LSD 0.05		0.9	1.5	1.5

The populations collected in Syria and Jordan showed the highest level of summer dormancy and also a very early regrowth in winter; both these characters are highly desirable for adaptation. Among the populations collected in Syria, growth patterns were influenced by the average rainfall of the collection site (Table 3).

TABLE 3

SUMMER DORMANCY (0 = DORMANT - 5 = GREEN) AND WINTER REGROWTH (0 = MIN - 5 = MAX) OF 12 POPULATIONS COLLECTED IN SYRIA AND GROUPED ACCORDING TO ANNUAL RAINFALL OF THE COLLECTION SITE

Rainfall	No. of populations	Summer dormancy	Winter regrowth 12.1.81	Winter regrowth 28.1.81
450 mm or less	7	0.6	3.9	4.6
700 mm or more	5	1.9	1.8	3.4
LSD 0.05		1.2	0.8	1.1

It seems obvious that the populations collected in sites with 450 mm rainfall or less, deserve consideration for both the ability to survive the hot and dry summer of northern Syria and to regrow actively in winter.

The seed multiplication of this material is, at this stage, the main bottleneck for further exploitation. The method currently adopted to produce seed for replicated trials makes use of pollination bags placed shortly after ear emergence on the same plots used for evaluation. A piece of cotton wool is placed at the bottom of the bag to hold shattered seeds.

Initially a metal rod was used to hold the bag; this was later discontinued because a strong wind can break the bag due to the pressure against the rod. Two types of bags currently used are the terylene bags described by Foster (1968) and common paper bags.

Although it was not our intention to determine the efficacy of those types of bags, our experience to date would suggest that the common paper bags are better suited for our type of environment.

The production of seed for long term storage of those accessions which do not have immediate value poses the same type of problems already mentioned. Usually 5 m rows and/or plants grown in pots are used to produce the small amount of seed (5 - 10 g) which is placed in sealed plastic bags for long term conservation.

Multiplication of the most promising accessions are used for large scale evaluation.

The production of adequate amounts of seed required for yield trials, regional testing and on-farm trials, is now a major problem restraining the expansion of the Forage Program at ICARDA. Large scale testing of the most promising material selected from the breeding and evaluation trials is necessary. The major constraints which limit a substantial and rapid increase in seed production are climatic factors and the lack of staff specialised in seed multiplication techniques.

Climatic factors in northern Syria do not favour seed production because at the end of the rainy season, temperatures rise abruptly. A combination of high temperatures and strong winds lead to rapid dessication of the above-ground vegetation and to a rapid loss of moisture from the developing seeds; furthermore, severe losses are sustained through shattering if harvesting is delayed. Climatic conditions pose serious problems for effective isolation. In the case of forage grasses for which isolation is spatial, strong winds during flowering impose a restriction on the number of entries which can be multiplied in a given area since isolation distances must be considerable in order to avoid pollen contamination. For alfalfa and other cross-pollinated legumes, isolation is achieved by means of pollination cages. Here also the high temperatures and strong winds tend to affect and disturb the activities of bees and hence effective seed-set.

The availability of trained staff, specialised in seed multiplication, is an important constraint. Within the present framework of the programme research activities, staff responsible for the appraisal and testing of materials are also responsible for seed increase. It is known that for forage crops, agronomic practices required for forage production are quite different from those required for seed production. This is not the case for grain crops where grain production is in fact seed production.

For meaningful expansion of the Forage Program activities, the establishment of a seed multiplication unit would play a key role. Such a unit has been established at CIAT (International Centre for Tropical Agriculture, Cali, Colombia) where it focuses on seed multiplication problems and acts as a catalyst both for the training of competent seed technologists and for the increase of seed supplies of potentially good lines.

Seeds are considered a very effective vehicle for conveying the benefits of improved crop technology to the farmers and for translating crop improvement research to actual production on farmers' fields. Increased production through improved varieties can only be attained if the seeds are available to farmers. The solution of seed production problems is therefore vital and deserves high priority at all times. At ICARDA, failure to achieve a rapid solution of some of our immediate problems of seed multiplication may very well delay progress and hence have an early impact on forage production in the region.

REFERENCES

Allard, R.W., 1960. Principles of Plant Breeding. John Wiley and Sons., Inc. New York, London.

Federer, W.T., 1956. Augmented designs. Hawaiian Planters. Record 55, 191-208.

Foster, C.A., 1968. Ryegrass hybridisation: The effect of artificial isolation materials on seed yield and floral environment. Euphytica 17, 102-109.

DISCUSSION

J.T. Williams *(FAO)* This programme illustrates something which is not widely appreciated, particularly by bureaucrats, and that is that it is frequently necessary to have an introduction programme. Institutes tend to introduce large numbers of seed samples and then it becomes immediately painfully obvious that it is necessary to go and collect local population samples and then rethink. A problem does arise then with the maintenance of these vast numbers of samples. The ICARDA list is a typical example; it probably needs a lot of rationalisation and it may be that many of the samples will have to be reduced because they are only sub-samples of sub-samples of something else from some other institute. To maintain, multiply and regenerate such samples may not be worth the effort; it is better to go back to the original institute or come to some sort of co-operative arrangement with them.

Henny Roelofsen *(Netherlands)* Dr. Somaroo, you mentioned waiting nine months for equipment. If it is any comfort to you, this is a general problem. Some people here in Europe have to wait for years.

B.H. Somaroo *(Syria)* Yes, but I was talking about nine months after you have ordered the equipment and paid for it! I did not have such problems when I worked in Canada.

W. Plarre *(Federal Republic of Germany)* Do you have a special programme financed by the EC? I know that the Community does finance special programmes in the international institutes. I would suggest that this meeting should recommend that there is a necessity to support your work especially in collecting primitive types, and that financial support should be forthcoming from the EC.

J.T. Williams Perhaps I could comment on that because I know the international centre set-up very well. I am not sure whether the EC does give core funding to ICARDA for its forage programme - I have not got this programme budget with me - but I know there have been discussions for the establishment of a genebank. ICARDA has had various propositions. One was that the genebank should be located in some buildings towards the north of Lebanon but the political situation has made this difficult. Now, plans have been drawn up and it looks as if it will be very near the Tel Hadya station in Aleppo; the governing board has given its blessing to the project. The IBPGR is also involved but, of course, it does require capital funds which other institutes cannot give it. It would certainly be very useful if the EC did express interest in this programme.

I would say also that when it comes to the collecting work, this has been taken by ICARDA as an essential remit of its terms of reference. It is taking some time to establish ICARDA and get it running. As you know, originally it was going to have a sub-station in Iran, it was then going to have a sub-station in Turkey, and the latest move as far as I know, is that the sub-station will be in Morocco. There have been a lot of difficulties; you can see the magnitude of the task. However, with the type of material they are going to have it is almost essential that they have storage facilities available because otherwise the material that is introduced or collected will just degenerate. We have been involved with ICARDA; I have an assistant there who is assisting on germplasm work but primarily for field crops rather than for forages. The other help we have been able to give is through a junior trainee who is attached to ICARDA, assisting with the collection work. We have also helped in the northern Yemen. This is a programme that has to expand and it would be very useful if this can be recognised.

J.G. Hawkes *(UK)* I note that you have some 2 500 accessions of *Vicia*, Dr. Somaroo. I don't know whether most of those are *Vicia faba* or whether there is a wide range of other forage species.

B.H. Somaroo They do not include *Vicia faba*; that species is being worked on by the grain legumes programme. These are a number of other species. At the moment we are concentrating on *Vicia sativa* and *Dasycarpa*. For the time being we have to narrow down the species to the ones that we think have the most potential.

J.G. Hawkes I wanted to ask how much of the *Vicia faba* material was collected in the Near East and how much came from European genebanks.

B.H. Somaroo I think a substantial amount of that material came from European genebanks; there was very little from the Near East.

J.T. Williams Perhaps as Chairman I could amplify a point about the need for training because an example was given about one of the North African countries growing everything all together. I can give an even more vivid example where for several years we were involved with another North African country to collect *Medicagos*. Of course, sub-samples of the material were left in the country and we tried to train people in collecting and thinking whilst these missions were going on and that national programme itself also grew its material in the same way. So it was virtually useless at the end, in terms of characterising and keeping the populations. It may seem a very simple task to many of us that there has to be isolation or there has to be some practical way of growing the material out, but it is not frequently followed.

H.A. Cortessi *(Greece)* Dr. Somaroo, can you clarify for me exactly what is meant in your Table 1 by 'Total no. of accessions'?

B.H. Somaroo It is the total number of accessions from Europe, Africa and whatever we have collected by our own effort.

H.A. Cortessi And what is 'Accessions acquired'?

B.H. Somaroo 'Acquired' is what has been given to us by other institutes.

Natural selection in populations of cross-pollinated species

H.L.PETERSEN
Royal Veterinary & Agricultural University, Copenhagen, Denmark

INTRODUCTION

One of the tasks of a genebank is to preserve germplasm collections with minimal loss of genetic integrity. However, from time to time seed regeneration is necessary. In cross-pollinated species each plant in a population will be genetically different from all other plants in the same population and different from the plants in the next generation. In order to preserve such populations during seed regeneration the gene frequencies of each character should remain unchanged from generation to generation, i.e. natural selection pressure should be minimised.

It might be difficult or even practically impossible for a genebank to propagate each seed lot in its natural habitat or even in a similar one. Seed regeneration outside the normal habitat of a population will lead to natural selection and hence to more or less severe changes in genetic composition. Genetic diversity in a variety thus makes the maintenance of genetic constancy difficult. Natural selection will be successful in increasing the proportion of types well-adapted to the habitat for seed production. Types that produce abundant seeds will increase whilst alleles that contribute to low seed yield will decrease. Under different environments, natural selection will develop different gene pools. Populations of out-breeders tend to move towards an equilibrium when propagated in the same habitat for several generations. When the habitat is changed gene frequencies will shift towards a new equilibrium.

In this paper some changes caused by natural selection during regeneration are outlined.

Figure 1 describes the main factors determining seed productivity. The model starts with seeds produced by a given population of a cross-pollinated species such as *Lolium perenne*

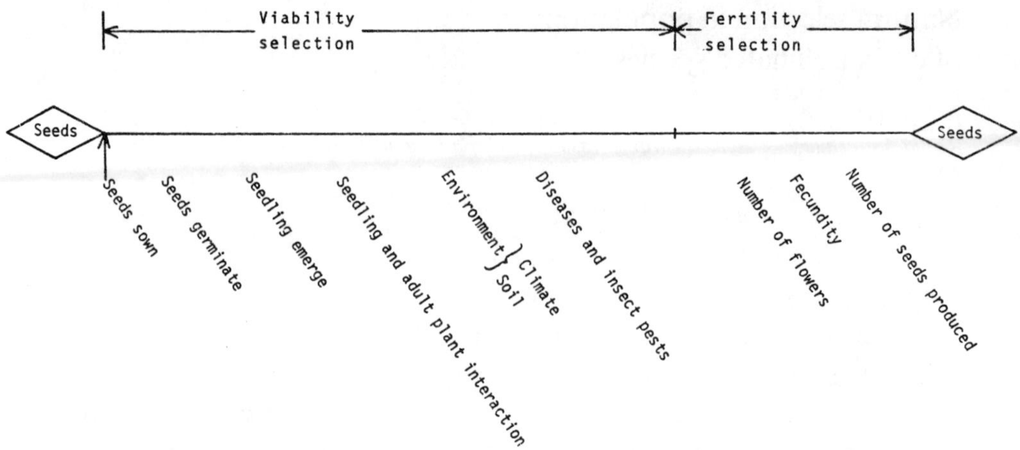

Fig. 1. Natural selection may occur in different stages of the life cycle.

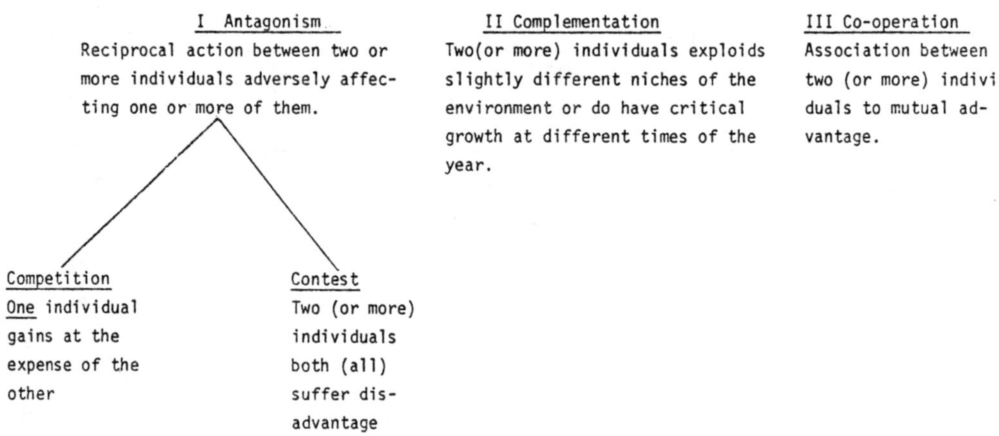

Fig. 2. Interactions between plants in close stands (Mather, 1969; Jacquard, 1973).

or *Trifolium pratense*. Seedlings emerge from stored seeds and start to compete in order to survive. The adult plant population will be a result of the influence of environment on the genotypes. Some, but not all genotypes will flower and set seeds. During the course of these events the population may change drastically especially when grown outside its normal habitat.

FROM SEED TO SEEDLING

The development of a seed to a seedling depends on the environmental conditions prevailing in the immediate vicinity of the seed. Seen from the point of the seed the environment is in most cases very heterogeneous. A random stone can cause late emergence, early competition from weeds can reduce the number of cultivated plants. Extra water supply may give an early start etc.

Genetic differences in germination capacity and ability to survive the first few days are of importance.

CHANGES IN JUVENILE AND ADULT POPULATIONS

In a sward a given plant will always be under the influence of its neighbours. Different categories of neighbour effect are defined in Figure 2.

Under normal conditions for seed production competition will probably be prevalent. Competition takes place when plants make demands on water, light and nutrients in excess of the supply.

The presence of a plant changes the environment of its neighbours and may alter their growth rate and form. Higher plants can respond to competitive stresses either by varying the death rate of individuals or the rate of development of leaves, tillers, flowers and seeds, etc., in the individual plant. When exposed to severe competitive stress some genotypes will produce a fairly high number of seeds, others less, and many genotypes will die before seed setting.

Charles (1961) demonstrated the effect of natural selection in the juvenile and adult stage (Figures 3 and 4) in

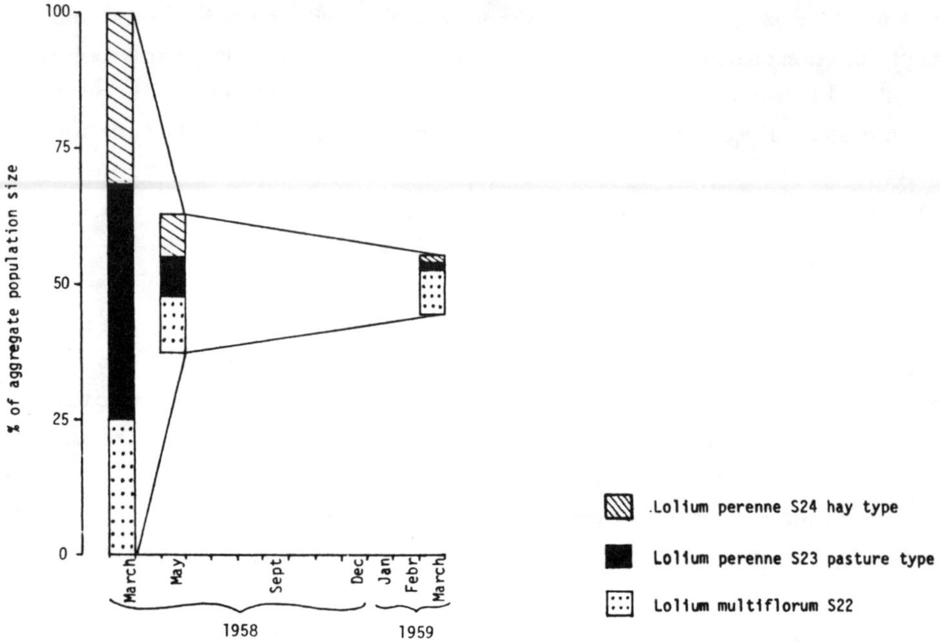

Fig. 3. Changes in *Lolium* population size - Charles (1961).

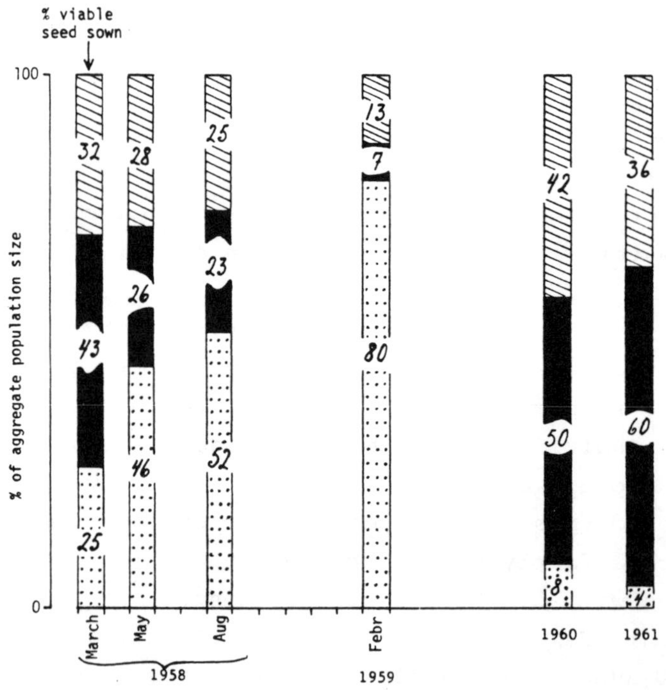

Fig. 4. Changes with time in the proportion of sub-populations of *Lolium*. (Charles 1961 and 1964).

a *Lolium* mixture comprising 9 kg/ha of each of the following varieties:

a) *Lolium multiflorum* (S22) from ordinary commercial stock selected for persistency and leafiness.

b) Early-flowering *Lolium perenne* (S24) bred from material collected from lightly-grazed pastures.

c) Late-flowering *Lolium perenne* (S23) based mainly on material from very old grazed pastures.

Plants were removed from each plot at intervals of 2, 5 and 11 months after sowing, grown on as single-plant units in a split-plot randomised-block, and classified into the cultivars. Differences in growth form between cultivars were readily apparent from the check rows of the three cultivars randomly distributed in the layout, where they could constantly be referred to at all stages of growth and in all seasons. Classification of growth form was supplemented by data on head emergence and plant height.

The pronounced decrease observed in total population size during establishment is illustrated in Figure 3; 25% of the viable seeds sown produced plants which survived two months after sowing, whereas only 11% survived the first winter in the sward.

The size of different sub-groups within the *Lolium* population did not decrease to the same extent (Figure 4). The rapid-establishing S22 dominated the swards, contributing 46% shortly after sowing, but 80% nine months later. Both the early and late-flowering *Lolium perenne* contributed less to the total population as the sward became older, the mean recorded figure for S24 was 13% and that for S23 only 7%.

Lolium multiflorum decreased from a mean figure of 80% to 8% in 1960 and nearly died out in August of the third harvest year. In contrast there was a rapid increase in *Lolium perenne* especially the S23 pasture type.

These experiments and many others suggest the following general conclusions:

1. One of the responses shown by plants to competitive stress is a reduction in seed output per individual.

2. In some species mortality rates among juvenile plants are very high.
3. The probability of death decreases as the individual increases in size. Larger individuals not only withstand stress for longer periods but also display greater competitive abilities.

WINTER INJURY

The selective effect of differential winter injury may be illustrated by seed production of an alfalfa variety outside its natural habitat.

Seeds of the alfalfa variety Ranger were produced in 16 regions of North America representing the states of Montana, Washington, California, Arizona and Mexico (Smith, 1955). Foundation, registered or certified seeds from these regions were sown in Wisconsin in May and cut back uniformly to the crown on September 10 1953. The following spring, winter injury estimates were made in mid-April (Table 1). No significant differences in performance were found among the plant populations derived from seed lots of Ranger of the same class (Foundation, registered or certified) of seed produced in the same state or produced from different ages of stand in the same state. An exception was the difference in winter injury between two second generation increases from Arizona.

The Ranger plant populations from seed originating from one generation of increase in Montana and Washington (Mont. foundation and Wash. registered) had only a small amount of winter injury. Plant populations from seed originating from one generation of increase farther south (Cal., Ariz., and Mex. certified) had incurred more winter injury. These tendencies were more apparent in the populations derived from Arizona seed than from California seed and were still more pronounced in the populations derived from Mexico seed.

The performance of the Ranger plant populations from seed originating from two generations of increase in Montana and in Arizona was strikingly different. The degree of winter injury in the second generation increases from Montana (Mont. certified) was virtually the same as that in the first generation increases

TABLE 1

WINTER INJURY IN SEED LOTS OF RANGER ALFALFA GROWN AS INDIVIDUAL PLANTS (Smith,1955)

State or country in which seed class[x] was produced		Generations in state or country	Number of seed lots tested	Seed produced from[x]	Percentage winter injury
Montana	F	1	2	Nebr.F §	8
Washington	R	1	2	Mont.F	7
California	C	1	3	Mont.F	13
Arizona	C	1	3	Mont.F	19
Mexico†	C	1	2	Mont.F	23
Montana	C	2	2	Mont.F §	7
Arizona‡		2	2	Ariz.C	31

Significance among population means for origins of Ranger seed using the Duncan 5% multiple range test

	Mont.F.	Wash.R.	Mont.C.	Cal.C.	Ariz.C.	Mex.C.	Ariz.‡
Percentage winter injury:							

x F = Foundation seed, R = Registered seed, C = Certified seed.

§ Information not available for one of the seed lots.

‡ Second generation of increase in Arizona and not eligible for certification.

† Seed produced by the Rockefeller Foundation at Mexico City, Mexico, under the 1950 standards of the International Crop Improvement Association for production of certified seed.

from Montana (Mont. foundation). In contrast, more winter injury was noted in plant populations from seed after two generations of increase in Arizona than in the Arizona first generation increases (Ariz. certified).

Sylven (1937), working with white clover *(Trifolium repens)*, red clover *(T. pratense)*, timothy *(Phleum pratense)* and meadow fescue *(Festuca pratensis)* in Sweden, demonstrated the danger of growing varieties of these plants for seed outside their native growing districts. This is especially the case if it is desired to maintain and further to fix varietal characteristics that are adapted to the climatic conditions of the respective districts. Seed for propagation fields should always consist of controlled elite seed supplied by the respective breeding station or its vicinity.

This experiment and many others suggest the conclusion that winter injury may cause natural selection.

SEED PRODUCTION AT DIFFERENT LOCATIONS

Stanford et al. (1960) and Laude and Stanford (1960) demonstrated genetic shift in ladino clover which is a cross-pollinated forage crop. They planted three identical polycross nurseries at three different locations with different climates (Table 2). In each polycross the 21 original clones of the variety Pilgrim were replicated 21 times and completely randomised. Synthetics from these locations were grown at Davis, California. About 500 plants from each seed source were examined and classified for the presence of the following characters:

1) red flecking of leaves, produced by a dominant gene in one clone;
2) a hydrolysing enzyme, linamarase, controlled by a dominant gene present in one clone;
3) cyanogenetic glucosides, also controlled by a dominant gene, and present in six clones; and
4) a white V-mark in the leaf, which was present in all but one clone.

The high frequency of red fleck in plants obtained from seed grown at Prosser was not unexpected because the red-flecked clone was rated second highest in vegetative vigour and flower production. At Aberdeen this clone had been below average in floral production; at Davis it was ranked eighteenth in floral development.

At Davis the clone producing the enzyme linamarase was very productive, but showed below average productivity at Prosser and Aberdeen; these rankings are reflected in the frequencies of offspring possessing the enzyme.

TABLE 2

EFFECT OF LOCATION ON THE COMPOSITION OF PILGRIM, A VARIETY OF LADINO CLOVER. (Laude et al., 1960 and Stanford et al., 1960).

Location	Latitude, Deg. N	Mean temp. of January	July	Mean frost-free season, days	Percentage of plants with classified characteristics Red fleck	Enzyme	Glucoside	Petal V-mark
Davis, California	38	45.4	75.3	242	0.02	21.45	19.79	86.27
Aberdeen, Idaho	43	19.8	69.9	102	2.69	2.98	12.38	90.22
Prosser, Washington	46	31.0	72.2	169	18.12	4.08	9.68	85.09
% expected					4.76	9.52	26.87	86.31

The clones possessing the glucoside produced less than expected at all three locations, particularly at Aberdeen and Prosser. The frequency of plants with V-mark was as expected at all locations.

These data clearly show that environment can have profound effects upon the composition of an open-pollinated variety even during one generation.

SEED PRODUCTION BY GENOTYPES AT DIFFERENT LOCATIONS

Clones of the above ladino clover variety were compared at three locations in terms of the percentage of the total seed harvest contributed by each. The results are shown in Table 3.

Most variation among the clones occurred at Davis, less at Tulelake. Progressively in moving north, the higher yielding clones produced relatively less and the lower yielding clones more. This increased uniformity in the production of the clones indicates a reduced shift in the composition of the seed harvested at the more northern of the experimental sites. The environment at this northern location resembles that of the areas of original clone selection and synthesis. It thus appears that the disproportionate seed contribution of the components of a synthetic variety is greater with increasing difference between the environment of the region of parent clone evaluation and synthesis and that of seed production.

TABLE 3

PERCENTAGE OF TOTAL HARVESTED SEED CONTRIBUTED BY CLONES OF LADINO CLOVER AT DAVIS, TULELAKE, AND PROSSER (Laude and Stanford, 1960).

	Location		
Percentage contributed by:	Davis	Tulelake	Prosser
	%	%	%
Highest yielding clone	22.3	12.9	8.6
Lowest yielding clone	0.0	0.0	0.0
Second lowest yielding clone	0.1	0.9	2.1
Highest 2 clones	41.4	23.3	16.5
Highest 4 clones	64.0	41.0	31.8
Highest 7 clones	80.0	61.0	51.9
Lowest 7 clones	5.0	9.1	15.0

SEED PRODUCTION AT DIFFERENT DATES OF HARVEST AS DEMONSTRATED BY CLONES OF LADINO CLOVER

The influence of date of harvest in the contribution of each of the ladino clones to the seed lot harvested is illustrated in Figure 5.

Since flowering commenced in late April, the July harvest contained roughly that seed produced in the first half of the season, and the October harvest the seed produced in the second half. Environmental conditions prior to each harvest differed considerably and in a sense the results indicated the response of the same plants to different environments. Prior to the July harvest, both temperature and day length were increasing while later in the summer temperatures were higher and day lengths were decreasing.

The relative production of a given genotype at one location may be altered distinctly by date of harvest.

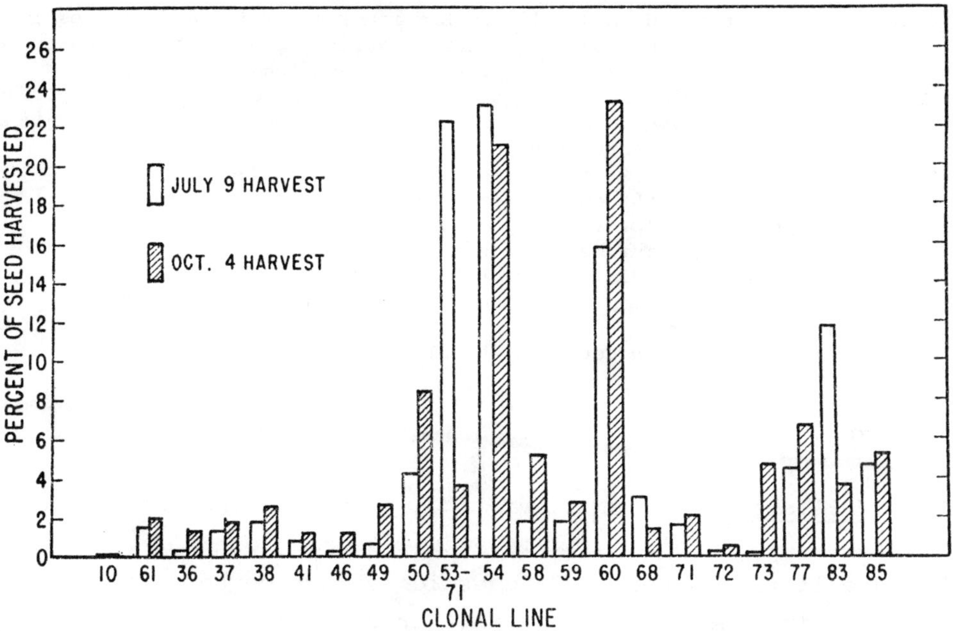

Fig. 5. Percentage of harvested seed contributed by each clonal line of ladino clover at Davis in 1957. (Laude and Stanford, 1960).

CONCLUSION

The experiments described exemplify the influence of natural selection in operation in cross-pollinated populations during one generation. The more the locality of seed regeneration deviates from the natural habitat of the population, the greater the risk for changes in population characteristics. This risk will be accentuated by several generations of regeneration in a deviant environment. When the aim is to preserve varieties of cross-pollinated species, propagation should if possible be undertaken in the natural habitat of the particular variety.

REFERENCES

Charles, A.H., 1961. Differential survival of cultivars of *Lolium, Dactylis* and *Phleum*. J. Brit. Grassl. Soc. 16, 69-75.

Charles, A.H., 1964. Differential survival of plant types in swards. J. Brit. Grassl. Soc. 19, 198-204.

Jacquard, P., 1973. Glossary of terms and definitions, 6th issue. Eucarpia Fodder Crop Section, 32 p.

Laude, H.M. and Stanford, E.H., 1960. Environmentally induced changes in gene frequency in a synthetic forage variety grown outside the region of adaptation. Proc. 8. Interna. Grassland Congress: 180-184.
Mather, K., 1969. Selection through competition. Heredity 25, 529-540.
Smith, D., 1955. Influence of area of seed production on the performance of Ranger alfalfa. Agr. J. 47, 201-205.
Stanford, E.H., Laude, H.M. and Enloe, J.A., 1960. Effect of harvest dates and location on the genetic composition of the synI generation of Pilgrim ladino clover. Agr. J. 52, 149-152.
Sylvén, N., 1937. The influence of climatic conditions on the type composition. Imp. Bur. Plt. Genet. Herbage Bull. 21.
Valle, O., 1960. Experience with seed production of northern European forage crops grown at various latitudes in North America. - Proc. 8. Interna. Grassland Congress: 185-188.

DISCUSSION

J.T. Williams *(FAO)* That was a very interesting paper Dr. Petersen. Your conclusion is an interesting one for discussion because it raises the whole question of how conservative conservation ought to be, or whether we ought to be interested in preserving genes with a certain number of adaptive complexes. It seems to me that there are great difficulties in trying to do this work in the natural habitat because it is very difficult to define the natural habitat. If you have a mosaic of ecological niches, as in the work of Bradshaw and so on, where even a few centimetres changes the habitat grossly in relation to selection pressure and changes, I am not sure how you would cope with it. However, it is an interesting field for discussion and it also relates very much to what Professor Demarly was saying this morning about the need, in some cases, for some *in situ* conservation.

H.L. Petersen *(Denmark)* I am thinking primarily of agricultural varieties rather than of natural samples. Therefore, I think it is best if a variety bred for Danish conditions is propagated fairly close to this area - not in Bari in southern Italy. That is the point I am making.

E.L. Breese *(UK)* I certainly agree with that last comment so far as the bred varieties are concerned, that under the conditions of competition which you outlined, they should be bred as near as possible to the area of utilisation for which they are intended.

Two points emerge very forcefully from your presentation, particularly from the work of Charles and others, and these are the effects of competition and the density aspects which end up with a 90% elimination in Charles' experiments after just nine months overwintering. It was those kinds of considerations which drove us, in Aberystwyth, into the controlled environment and the controlled plant density situations of the pollination chamber.

There is one further point here, as Dr. Williams said, how do you define a habitat and how then do you regenerate by seed from it, if we take the Bradshavian case or indeed the small localised population? They are there chiefly due to their survival by a sexual selection through vegetative propagation so if you take just one generation, either there or elsewhere, through seed, you will inevitably get a swing back, even under isolation, to selection of those which produce most seeds again. Now, in natural situations that contamination resulting from pollen flow is kept in check, as Bradshaw points out, by continuing forceful selection. When we are preserving a particular combination, however we do it, if we can't keep the natural habitat going, what we want to do is to lift the plants and try to freeze them. You illustrate so many of the points we have tried to take into account: low density, try to get every plant represented of the sample we first took, in the way that we have suggested. You have put that competitional aspect so forcefully, that antagonism, and the elimination.

M.H. Poulsen *(Denmark)* I am now thinking as a practical breeder in terms of the multiplication we do, in connection with the results we have seen here, and with the change in the juvenile stage of sowing out the cultivars and the changes which occur, the losses, and I would like to ask, in relation to what Dr. Breese has said, what knowledge do we have about the genetic constitution of the surviving plants after these nine months? Do they still represent the original cultivar? If they do that, then we have not lost anything; if they do not, we would expect to see what was also shown, a tremendous shift towards something else. As far as I know S23 has not changed that much during the multiplications.

H.L. Petersen Some people say that this very strong competition will kill the weakest. The other problem is that if your normal method of propagating your varieties is just by seeding in a sward or in very strong competition, and you have had an equilibrium in your variety by sowing and getting your seed in that way, I think you should continue to preserve the variety in the same way. I understand the difference in your view and that of Dr. Breese. He tries to get rid of that competition as far as I understand it. Is that right Dr. Breese?

E.L. Breese For preserving the populations that we collect for, shall we say, posterity.

H.L. Petersen The difference of opinion might be just a question of in which way do you try to preserve your variety. Some time ago in Aberystwyth, I know that you tried to preserve your varieties of grasses by preserving your clones. That is probably right. However, in most cases in Denmark, you have your breeder seed and it is propagated without keeping the clones, so it is a matter of opinion as to which is the best way to preserve a given variety. It depends on the way in which you try to keep your variety - to a certain degree at least.

E.L. Breese Ideally we would wish to maintain the basic plants so that we can limit the number of generations of increase, because for every generation of increase all the risks that we outlined this morning, of shift, of drift, of isolation, will be there. That is why we require to keep the basic plant. We have also had difficulty in doing that in the past because of random loss from virus etc. of the basic plants. That is why we have gone to tissue culture. But still the principles thereafter come much more in the context of the rules of seed multiplication which are very important within the district, within the variety that was formed, or in analagous districts with analagous rules of the timing when you cut for heading - this is very important too. There are so many rules to observe and one has to go along with that.

I should add one more point here since I am familiar with the experiments of Charles, my colleague. He did go further in the trials you referred to and he tried to look at shift, at actual change in the varieties themselves, through the succeeding years. This is a very difficult thing to do. Since they are dynamic populations, even as they are maintained in the vegetative state in the more permanent pastures, they do change. Sometimes in breeding, we have put a particularly hard management on them for many years and then taken survivor seed to give us a new population.

J. Jakubiec *(Poland)* As a breeder, would you prefer to maintain a gene pool with many hundreds of genotypes in it, or subdivisions of the gene pool as ecotypes?

H.L. Petersen As a breeder, one is very interested in genetic diversity and in that case I would prefer to have a gene pool where you cross all together and then make your selection out of this. It is something like the composite crosses in barley.

W. Plarre *(Federal Republic of Germany)* With regard to this last example that you have given, it depends on the breeding goal. It could be completely different; if you are looking for late barley, for example, then you cannot always expect the greatest diversity. You also need to know the special environmental conditions in which really late barley has the best capacity for developing. Then you have also to change the environmental conditions. I think we have different methods, from the practical breeding point of view we have to look for these special environmental conditions. On the other hand, if the aim is simply to maintain our gene resources then we have to look at other factors. We must differentiate here.

H.L. Petersen Yes, I agree, but I was thinking as a breeder. For instance, if you are going to breed some varieties of *Lolium*, you might have a very diversified gene pool and that gene pool might just be propagated for a number of years in order to let the natural selection adapt the pool to the environment for which you want to breed. Of course, it will take some time but then you can select from that pool. The other way was just to keep these ecotypes by themselves, multiplying just by pollinating inside the ecotype. In this way you are probably able to keep the ecotypes and select certain types. Later on it is possible that you can make your variety by crossing types from different ecotypes, that is another way of approaching the problem. It is really for the plant breeder to decide on the method he wants to use.

J.T. Williams I think this question of gene pools may well come up in the discussion again. There are some fundamental differences depending on the breeding aims; I think terms are used very loosely. There is a very big difference between racial composites as a gene pool which are based on genotypic factors, and gene pools based on mixtures of ecotypes which are probably frequently physiological responses and you can't see them by eye. So I think there are differences in concepts here which are sometimes blurred in the discussion.

E. Porceddu *(Italy)* When you transfer a population from one place to another, you can have plants that flower earlier than the majority, and others which flower later, and so on. Therefore, the mating is not at random but becomes assortative. What would be your opinion, Dr. Petersen, on sub-dividing the population into three different parts, the early, medium and late types, and preserving them separately, evaluating them together, but keeping them separate when you are regenerating the seeds? Would that be better than keeping the whole population multiplying at the same time in the same plot and taking the risk of making selections for early or late types?

H.L. Petersen If you divide a population into three different parts as you suggest, you change the competition at least. It might give you a change of variety also, even if you try to bring them together when you evaluate. Of course, I cannot answer the question fully.

Reproductive patterns of *Medicago sativa*

P.GUY
INRA, Lusignan, France

YVETTE DATTÉE
Université Paris XI, Orsay, France

INTRODUCTION

The cultivation of lucerne has been linked with the history of man for nearly 10 000 years. Different names have been given to this plant: *Medicago*, alfalfa and lucerne (laouzerdo, userdas, eruye, luzu, lizueu etc.). In France, the plant is called lucerne, the name given to it by the Catalonians in the 16th century. Lucerne, *Medicago sativa*, was first introduced into France during the Roman era by the Romans, and then, in the 8th century by the Arabs. It has flourished in France and throughout Europe since the Renaissance period.

Lucerne adapts well to new environments with hybrid traits between *Medicago sativa* and *Medicago falcata*. In France there are six common types of lucerne: Flamande, Ormelong, Poitou, Marais de Luçon, Marais de Challans and Provence. "Alfalfa is a naturally outcrossing perennial that depends upon bees for pollination. The flower is complete, therefore selfing can occur. In nature, however, the frequency of selfing is usually much less than that of crossing." (Busbice, 1972). We agree with this conclusion. However, significant variability in selfing rate, ranging from 5 to 50%, has been reported.

Allogamy (cross fertilisation) generally predominates, first because an insect is necessary to trip off the flowers, and secondly because of the selective advantage of foreign pollen. There is a strong relationship between the plants, the pollinators and the environmental conditions.

RELATIONSHIP BETWEEN THE ECOTYPE AND THE ENVIRONMENT

Ecotypes and variety populations are in a state of equilibrium with their environment because of the insect pollinators and the pedo-climate. Populations are characterised not only

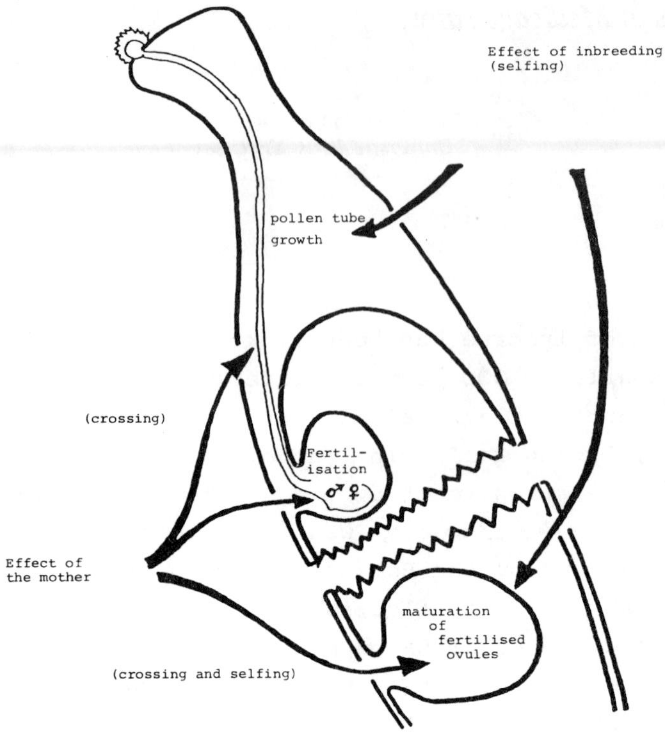

Fig. 1. Correlation between self- and cross-fertility of female parent. Common effect of female parent on progeny. (Dattée)

by their morphological and physiological features but also by their reproductive systems. The respective rates of self- and cross-fertilisation are particularly important in determining and establishing genotypical structures. Using lucerne, we are able to study the way plant and insect co-adapt within a given environment.

The pollinating insects

Composition

In France and in other parts of Europe the bee, *Apis mellifica*, which visits *Medicago sativa* mainly for nectar, plays a minor role in its pollination even though bees exist in great numbers. The following are among the more important pollinators: *Melitturga clavicornis*, *Andrena labialis*, *A. ovulata*. *A. Flavipes*,

Eucera longicornis, E. clypeata, Melitta leporina, Bombus terrestris and *B. agrorum*. The visit of an insect is indispensable to fertilisation except when self-tripping occurs, but this rarely happens.

The composition, abundance and species behaviour of the pollinators play an important role in the reproduction of lucerne.

Distribution

The population of each insect pollinated species varies with the geographical location (zone), the season and the year. There is a wide distribution of *Melitta leporina, Bombus terrestris, Andrena ovatula,* and *flavipes* in France. *Eucera clypeata* is meridional. *Melitturga clavicornis,* absent in the north, is localised in the south and found in areas containing certain soil types. *Bombus agrorum* is found at the edges of forests.

Seasonal activity depends upon the species. For example, in France we observe *Eucera tuberculata* and *Eucera longicornis* in June, *Eucera clypeata* and *Melitturga clavicornis* in July, and *Melitta leporina* and *Andrena ovatula labialis* from July to September. These different species pollinate the lucerne depending upon their periods of activity and the flowering dates.

Differences in the precocity of flowering exist, depending upon the ecotypes and the type of management; there is also variability between the genotypes of the same ecotype. In addition, during the flowering period, it is likely that several species will contribute to pollination. Another important factor to take into consideration concerning the respective contributions of the various pollinators is the abundance and synchronisation of the flowering of other plant species which attract the same insects *(Trifolium, Melilotus)*.

Abundance

The abundance of pollinating insects varies from 30 to 1 000 individual insects/ha and at times can reach an exceptional 2 500/ha. The bocage zones (mixed woodland and pastureland) are generally more favourable than the open fields.

Behaviour

Numerous elements in the behaviour of the pollinators condition the respective contributions, as male or as female, of the different gentoypes as well as the rate of self-fertilisation. These behaviour factors are generally related either to the environmental conditions or are characteristic of the pollinating species. Warm and sunny weather, in general, favours the activity of the pollinators, increases the rate of self-tripping, and enhances the fertility of the pollen. A heavy density of pollinators obliges them to change flowers (inflorescence) frequently, in search of flowers not yet tripped off. Genetic mixing is therefore amplified. The differences can be important; for example in July 1975, in a small isolated trial test in which there were from 2 000 to 4 000 *Melitta* /ha, there were never more than one or two flowers that had not been tripped per inflorescence! In the southern part of France the ratio of tripped flowers to those which can be visited is on average 30%. It was possible to bring this ratio up to 45% because of experimentation with *Megachile rotundata* (Delaude and Tasei, personal communication). Moreover, several authors (Pedersen and Bohart, 1953; Vansell and Todd, 1946; Boren, 1962 and Kauffeld, 1967) studied the clone preference which can be defined and characterised by the rates of visits and/or the different trippings. According to these authors, colour and concentration, as well as volume of nectar, all play a part. In our case, when we have demonstrated important clone preferences (Table 1), we have never been able to obtain repeatable measurements of the genetic variability for concentration and volume of the nectar. With regard to colour, it appears that even at a distance the bumble bees differentiate between clones which are of identical colour to the experimenter (Table 2).

The fertility of the pollen and dehiscence of the anthers are important factors attracting wild bees in the greenhouse as well as in the fields. However, these features do not affect *Apis mellifica* whose main concern is collecting nectar (Table 3). For example, the differences in attraction are very important between male-sterile clones and fertile clones, but these differences diminish when one compares not only one clone but a

TABLE 1

ATTRACTIVENESS OF CLONES EXPRESSED IN TERMS OF ARCSINE $\sqrt{x}$ x = TRIPPING RATE IN CAGES, LUSIGNAN.

	Colour of flowers				
	White	p^3	p^4 b	p^4 R	p^5 R
Apis	49	42	0	70	72
Andrena and *Eucera*	34	43	10	44	28
Melitta	36	58	24	63	44
Bombus	35	42	22	37	46
$\bar{x}$	39	46	14	54	48

p_0 to p_3 intensity of purple colours

b variegated

r red

TABLE 2

ATTRACTIVENESS FOR *Bombus* OF 4 CLONES OF *Medicago sativa* OF THE SAME COLOUR IN CAGES.

	A	B	C	D	
Number of tripped flowers	19	10	4	17	S 1 %
Number of visited but not tripped flowers	27	11	14	14	
Number of visited racemes (with tripping)	9	5	4	7	NS
Number of visited racemes (without tripping)	9	8	6	3	
Number of tripped flowers/ visited raceme	1.06	0.77	0.40	1.70	
Number of tripped flowers at a time/clone	3.17	2.30	2.0	4.25	

male-sterile population to a fertile male population. The differences decrease even when one compares commercial varieties of lucerne (EUCARPIA trial, *Medicago sativa*). Although there are still varietal differences in pollen fertility (50 to 80%), this factor is no longer a preponderant one for attraction.

TABLE 3

COMPARED ATTRACTIVENESS OF MS AND MAINTAINERS CLONES

	Pollinators* /m^2		Quantity of pollen	% pollen fertility	*Apis* m/m^2 rank	
C. 21-2	0.28	14	0	ε	14	male-sterile
M_S R12	0.33	13	0	0	13	
M_S R21	0.58	10	0	ε	2	
213	0.52	11	0	ε	1	
M_S T_{11}	0.87	8	0	0	11	
M_S 52	0.73	9	0	8	9	
Lu 18	0			low		
GIE	0.08			low		
I552	2.48	3	3.5	46	10	maintainers
A_{11}	0.89	7	2	33	5	
A_{11}x $F_{11(1)}$	0.90	6	4.5	69	3	
OR 215	1.29	5	1	20	8	
K 57	2.00	4	4.2	52	7	
B 1	0.45	12	0.5	61	4	
E S4	3.24	1	2	76	12	Control
Europe	3.16	2	2	83	6	

* Sum of 11 counts

Does mobility difference exist and does this mobility difference depend upon the species of the pollinators?

A priori, an insect that changes plants often, resulting in unimportant numbers of trippings per raceme, favours allogamy. We do not have quantitative information concerning this at the moment. Ecalle estimates, however, that *Megachile p.* and *Melitta l.* are more active than *Andrena ovulata*, *Halictus marchali* or the bumble bees. Although the mobility differences of the pollinators remain to be confirmed, we have seen that factors modifying the genetic mixing carried out by the pollinators are numerous. The factors adhere principally to the abundance of the pollinators and to the frequency of their visits from one genotype to another. One can ask what the consequences of these variations are upon the genotypical structure of the populations, and particularly on the rate of selfing.

The rates of selfing observed in *Medicago sativa* are extremely variable. The insect pollinator is necessary for cross-fertilisation as is some variation in the rate of self-fertilisation. The results obtained in Lusignan in cages (Table 4) show the differences between species. In addition, the insects also induce fertility differences resulting from open pollination as well as from selfing (Table 5). The honey bee appears to be one of the most efficient elements in reproduction by self-fertilisation (Tables 4 and 5). Under natural conditions, the density of variation of the pollinators probably explains the different self-fertilisation rates observed by numerous experimenters. In other respects, the clonal preference can lead to homogamy in other characteristics such as the composition of the nectar or the colouration of the flower. From this we can say that the concept of random crossings between individuals of a population (pamixie) is undoubtedly theoretical and that the biological reality of the crossing system is much more complex.

TABLE 4

EFFECT OF THE INSECT SPECIES ON THE SELFING RATE IN *Medicago sativa* IN CAGES. LUSIGNAN.

	Bombus terrestris	*Apis m.*	*Megachile p.*
Plant in pot	15.2	13.5	10.5
Detached racemes	13.0	10.5	9.5

The pedo-climatic environment

Zygote selection is made depending upon environmental conditions such as climate, soil and intra and interspecific competition. Selection concerns both the gene structure (allelic composition) and genotypic structure (plants from cross-fertilisation as well as plants from self-fertilisation). Demarly (1957) pointed out the existing relationships between ecotype origin, winter regrowth, and resistance to the cold. More generally speaking, there is a relationship between the original latitude of the ecotype and the length of day as well as winter temperatures. Mediterranean ecotypes (Gabès,

TABLE 5

FERTILITY OF *Medicago sativa* ACCORDING TO INSECT SPECIES

		Open pollination	Selfing
Pod/flower	*Bombus*	79	/
	Apis	70	26
	Megachile	72	15.3
	Hand pollination	89	52
Seed/pod	*Bombus*	2.8	/
	Apis	3.3	1.2
	Megachile	3.1	1.4
	Hand pollination	3.7	1.7
Thousand seed weight g	*Bombus*	2.5	/
	Apis	2.9	2.1
	Megachile	2.8	2.15
	Hand pollination	3.15	2.4

Selfing rate: *Bombus* 15.2 *Apis* 13.5 *Megachile* 10.5 Hand pollination 0

Taguira, Provence) are capable of growing in winter, whereas the Canadian varieties and Flamande, *Medicago falcata* species cross well only when the days are long. We can also cite the ecotype 'Marais de Luçon' and the Soviet ecotype adapted for temporary submersion. Curiously, the disease factor seems to have differentiated the ecotypes or the variety populations to a lesser extent. Are attacks by *Verticillium* and by *Ditylenchus* too aleatory?

The plant

At each stage of reproduction (Figures 2 and 3) there exists a genetic variability which, depending upon the environment, permits an orientation towards allogamy or autogamy:

1. aptitude for self-tripping;
2. aptitude of the growth of the pollen tube (self-pollination) in the style;
3. aptitude for the maturation of the ovules (self-fertilised). (Dattée, 1975; Abdelkefi, 1980).

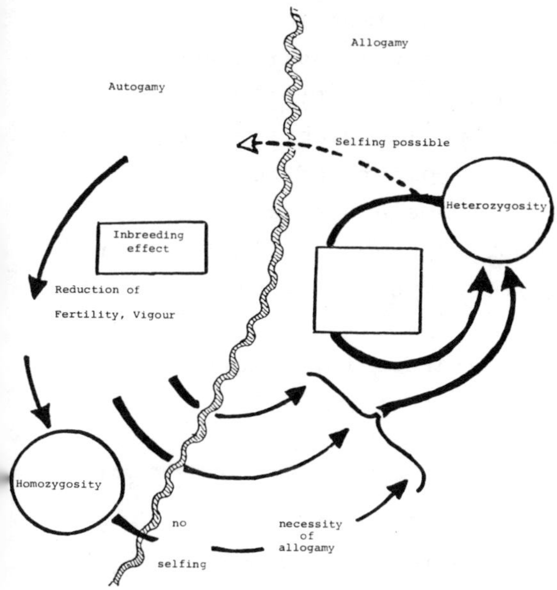

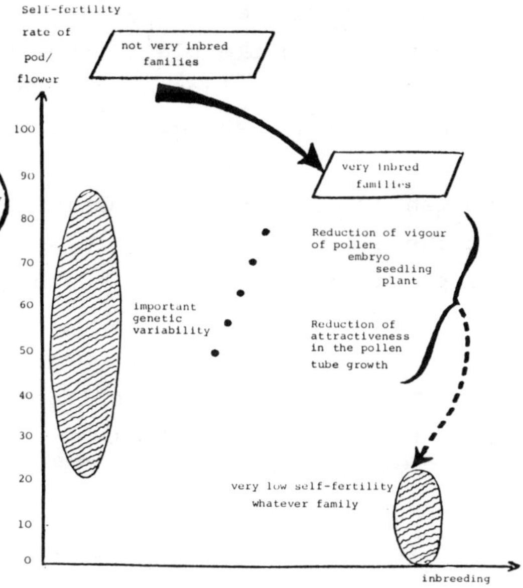

Fig. 2. Regulation of heterozygosity (Dattée)

Fig. 3. Self-fertility of lucerne according to inbreeding (Dattée)

Even though self-fertilisation is not conducive to seed production, there is no strict barrier to self-fertilisation. Tolerance to self-fertilisation displays genetic variation. Apart from results showing weak rates (from 5% to 15%) by Varga et al. (1973), Bolton (1962), Busbice (1972) and Guy et al. (1969), we find higher rates (from 30% to 50%) from Veronesi (personal communication), Kehr (1973) and Bober (personal communication) under certain conditions. It is difficult to question the differences given even if on the agricultural level it is correct to ask what the contribution of self-fertilising plants of successive generations is as far as number, weight and fertility are concerned. It is preferable to say that this rate of self-fertilisation varies with the conditions in the ecosystem and that it perhaps characterises not only the environment but also the lucerne population.

Temperature is also an essential element in the interaction between the plant, the insect and the environment. Pollen fertility increases with temperature. At 17°C numerous plants show a weak fertility, with a pollen viability between 5% and 25%. It appears, however, that plants of German and Polish origin are less sensitive to temperature variations. On the other hand, there is an important increase in the self-tripping rate in culture chambers at temperatures between 25°C and 30°C.

As we have already noted, temperature influences the activity of the insect pollinators, that of *Megachile pacifica* much more so than that of *Bombus* and *Apis mellifica*.

Other internal factors concerning the plant, (Figures 1, 2 and 3) probably characteristic of the population regulating self-fertilisation and cross-fertilisation, are genetic variability from So for self-fertilisation and self-tripping, the consanguinity inbreeding effect upon the vigour of the pollen (2n), the seedling, the plant and the effect of the relationship between the pollen and the ovules (stimulated or not from the growth of pollen tubes).

Do ecosystems exist in which the climate, the insect pollinator population and lucerne ecotype co-adapt and which possess particular characteristics of adaptation to self-fertilisation and cross-fertilisation? This question is important today and others have already given positive answers to it.

NECESSITY OF MAINTAINING VARIABILITY

Phenotypical and genotypical variability

There is no direct relationship between the extent of the genetic base and the phenotypical variability expressed. A population such as Flamande or Du Puits would have a variation coefficient for spaced plants from about 30 to 35% for weight per plant, close to that of a hybrid between two plants, whereas for certain morphological characteristics (colour of flowers, leaves, growth habit, etc.) the hybrid would be more homogeneous. A family may remain heterogeneous for several generations of self-fertilisation. Heterozygosity and polyploidy conceal the genetic variability (Benoit, personal communication) for vigour and enzymatic characteristics (number of electrophoresis bands). Biological interpretation is not always clear. While cross-fertilisation and heterosis maintain a special potential for variability expressed in terms of the number of isoactive alleles per locus, they buffer the phenotype to an average state co-adapted to the environment (law of greatest number, metastable state, resistance to fluctuation).

Self-fertilisation, homogamy, isolation of the ecosystems and eventual selective factors permit a realisation of the genetic variability of the species and at the same time result in a fertility loss of the population. It is remarkable to note that it is possible by simple mass selection (controlling the environmental effect) to extract important and varied tolerance to *Verticillium*, to *Colletotrichum* and to stem nematodes from a natural population such as Flamande. Although very sensitive to diseases and to crop pests, these populations conceal a hidden variability of different, multiple systems of resistance.

Variability within and among populations

Are the inter- and intrapopulation variabilities of the same nature? We admit implicitly that genetic distances are related to the heterosis potential. What is this? The distance between populations can be based upon several elements: the morphological character, quantitative or qualitative; the enzymatic character; and the host, predators and symbiose (virulence, biotype).

Reciprocal recurrent selection is an example of application. In a panmictic population, having a linkage equilibrium, the measurement of the distance from only a few markers (electrophoresis for example) theoretically gives very little or no information concerning the distance for the other parts of the genome, and hence, for possible heterosis on crossing. It is not certain that it would give other information concerning the distance between the families coming from a similar balanced population. However, they will probably serve as a useful index in measuring the distance between differentiated ecotypes (Dattée, 1975), which is also useful in filiation developments.

The enzymatic markers used in electrophoresis are still limited in number and unfortunately in nature. They represent only to an unimportant degree the entire knowledge of enzymatic functions. The observations of Brunel (1979) and Dattée (1975) point out a heterozygotic weakness in lucerne. Is this true for the entirety of the genome? One is able to, or must be able to, imagine that the variability within the interior of a population and between populations is not carried upon the same parts of the genome (in spite of inevitable overlapping).

Gene, linkat, general adaptation and reproductive behaviour

A population or an ecotype is a collection of genes, varying in frequencies, of which a certain number can be of agricultural or experimental interest: rare gene, vertical resistance (disease), specific gene (pest) maintaining male sterility, white flowers, markers, etc.

A population is a cytoplasm, a collection of linkats (a grouping of co-adapted functions), an original, complex architecture and an equilibrium which above and beyond elementary gene actions corresponds to a plant type, to a yielding aptitude and to a behaviour of the reproductive system.

The best evidence concerning the diversity of inter- and intrapopulation is the varied means of selection in order to arrive at the objective. The lucerne populations (very early) were obtained in two different ways: self-fertilisation (Demarly, 1957) and by natural selection of ecotypes (VIR Leningrad). Varieties of lucerne resistant to *Verticillium albo-atrum* were acquired in several ways, by recurrent mass selection (Sweden, Great Britain, France, etc.) and by interspecific crossing (GB) with natural ecotypes.

This demonstrates that the same phenotype can be realised in diversified genetic ways. At the present time, as lucerne is cross-fertilised, tetraploïd and still semi-wild, preservation may not seem to be essential. It must be noted however, that almost all the North African ecotypes having unique characteristics, namely the ability to grow and develop in short days, in moderate temperatures and possessing high pollen fertility, have been lost.

The plant breeder must preserve and build up genetic variability; many domestic plants no longer yield as well as their original populations. The plant breeder is interested in the genes, the functions and in all vegetative and reproductive behaviour which accompanies these features. He is interested in the plant ecosystem and the surroundings. Experimentally, the introduction of a new characteristic to a varietal creation often causes the modification of other agricultural characteristics which are unfortunately a disadvantage which must be overcome by a new selection in order to find the anterior characteristics. The examples are numerous: Texas cytoplasm in corn, rape free of erucic acid, corn of a high lysine content, palatable *Festuca arundinacea, Trifolium pratense* tetraploïd.

At the same time varieties of lucerne tolerating diseases (*Verticillium,* stem nematodes) rarely show a yield level similar to that of the parent plants during the first cycle of selection. One can therefore ask whether the answers correlated with other characteristics related to artificial selection do not depend upon the populations. If such is the case, the plant breeder must have at his disposition a choice of balanced genotypes and amongst these, depending on the objective of the selection, he will find variability in accordance with the importance of the correlated answers.

Population stability

Lucerne plants in the forms from Asia Minor have been cultivated in Western Europe and America since ancient times, becoming of more commercial importance during the Renaissance. There is still some mystery about the origin of lucerne and its polyploid nature. The role of its neighbouring pre-existing form *Medicago falcata* is not well known, although many botanists and plant breeders (Mayer et al., 1951) admit that there have been many crossings between new forms and the local ecotypes of *Medicago falcata.*

Within two to four centuries in France, (on virgin lands such as polders and the west marshes) and within one to two centuries in the Ukraine, variety populations were created and resulted in importantly differentiated phenotypes that we recognise even today. This gives us an idea of the rapidity of the evolution of the lucerne ecotypes and the risks involved if we were to change the agricultural ecosystem. Electrophoresis studies of genetic variability in isozymes show that in so far as the analysed loci represent the genome, the genetic variability of inter- and intrapopulations is of comparable importance. If we look at the complexity of the reproductive system and all the environmental factors which can modify the genotypical structure from one generation to the next, we can ask whether the system is a stable one and at the same time ask ourselves about the evolutionary rate. Paradoxically, the synthetic varieties which represent small populations appear stable if they are always produced with the original parents. This apparent contradiction is perhaps due to differing time scales. The number of generation multiplications arriving at a commercial seed from the constituents is low and on average the result obtained can be reproduced no matter what the climate conditions. In addition, the tetraploid confers a certain slowness in evolution. Long-term ecotype developments depend upon environmental variations. This continuation is not apparent from one generation to the next but the evolution is more flagrant if one compares it with many generations, that is over one or two centuries.

Maintaining variability

Preservation of genes and populations generally seems incoherent and anarchistic. At the present time, I believe that in Europe a list of the older generations does not exist. In spite of this, it would be hasty to believe that nothing has been done.

The VIR of Leningrad has certain types and the members of VIR participate in international studies. The Aberystwyth station has carried out systematic selections. The USA has an important number of selected parents. All plant breeders have varied, stabilised (clone) material or are progressing towards selection. The control and certification of seeds and the official list commercial varieties are the cause of the loss of great numbers of populations in some countries. At times, for example, in France, there are services which participate in the safeguarding of populations.

With regard to lucerne, genebanks exist which permit the maintenance of some simple genes such as white flowers and coloured grains. For the plant breeder it is often a question of monstrosities being useful as markers rather than as genetic resources.

Another type of bank exists whose objective is to preserve and maintain populations, variety and ecotypes. Nevertheless, the multiplication of small numbers outside their ecosystem does not permit preservation and/or stabilisation of these populations. We hope to be able to preserve at least an image, a fragment of the original populations. This permits the preservation of an important genetic variability potential requiring that the preserved and multiplied forms (not stabilised) be well chosen and numerous.

Recurrent populations, active selection or preservation concern improved artificial populations in which it is hoped to maintain for as long as possible variability which is useful to the plant breeder. It is a question of an agricultural objective, precise and definite, at all times, which is to preserve the maximum number of reproductive plants from each generation. However, although useful to the plant breeder, they are totally incapable of preserving the multiple behaviour characteristics of the species. On the other hand, we must make it clear that the plant breeder by simple or complex means (choice, interspecific crossing, mutation) builds up a new variability. Is it possible to compare today's species variabilities in apples with the variability of the original species? This does not make sense.

At ACVF (Association of Forage Variety Creators), the French plant breeders agreed to assure the maintenance (*in situ*) in the original regions of the French lucerne populations. These have been authentically protected by the SOC (Official Control Office). They are increased every five to ten years on a 1 000 m^2 surface in order to obtain 50 kg of seeds for general availability. It is a question of five populations: Flamande, Provence, Poitou, Marais de Challans and Marais de Luçon.

CONCLUSIONS

Genetic variability concerns genes, linkats, genome structure and the reproductive system. The common cultivated lucerne plants are in strong equilibrium with the ecosystem, agricultural techniques and environment.

All conservation methods necessitate sampling. There is no satisfactory preservation technique; it is necessary to combine several and if possible to control evolution by marking techniques such as electrophoresis and morphology.

Research should be carried out on the development of variability and the reproductive system within the species and on the balance between self- and cross-fertilisation, and also the new preservation techniques in *in vitro* culture. These results should then be compared with each other and with the traditional methods of preservation and of multiplication of seeds.

It seems that priority should be given to the older populations. Finally, it is felt that for lucerne populations which have evolved and are maintained by selection pressure brought about by agricultural systems, due consideration should be given to the extent that conservation may be obtained *in situ*, in addition to long term seed (or vegetative) storage.

The role of the plant breeder is not only to preserve but also to build up variability necessary to that building up. We must also be able to use the most varied and most balanced genetic resources.

REFERENCES

Abdelkefi, A., 1980. Technique d'étude en semis *in situ* des relations style-pollen chez la luzerne (*Medicago sativa* L.). Ann. Amélior. Plant., 30, 271-284.

Bolton, J.L., 1962. Alfalfa Botany, cultivation and utilisation. Leonard Hill, London, 474 p.

Boren, R.B. et al., 1962. Foraging behaviour of honey bees on selected alfalfa clones. Crop Sci., 2(2), 185-188.

Brunel, D., 1979. Recherche du déterminisme génétique de quelques systèmes enzymatiques chez la luzerne (*Medicago sativa* L.) di et tétraploïde. Th. de doc. 46 p. 3ème cycle Orsay.

Busbice, T.H., 1972. In: C.H. Hanson. Alfalfa Science and Technology. Genetics and Breeding Procedures 283-318.

Dattée, Y., 1975. Facteurs génétiques liés à l'auto - et à l'interfertilité chez *Medicago sativa* L. Th. Doct. Paris 208 p.

Demarly, Y., 1957. Biologie et exploitation de la luzerne. Ann. Amélior. Pl., 7(3), 247-272.
Demarly, Y., 1963. Génétique des tétraploïdes et amélioration des plantes. Ann. Amélior. Plantes 13, 307-400.
Guy, P., Lecomte, J. and Ecalle, C., 1969. Utilisation d'insectes pour la pollinisation de la luzerne en enceintes fermées. Ann. Amélior. Plantes 19, 333-343.
Kauffeld, N.M., 1967. Clonal preference factors related to honey bee foraging on alfalfa. Kansas State Univer. Ph.D. Thesis 67-13, 954.
Kehr, W.R., 1973. Cross-fertilization of Alfalfa as affected by genetic markers planting methods, locations and pollinator species. Crop Sci. 13, 296-298.
Mayer, P., Vincent, A. and Ecochard, R., 1951. Les populations françaises de luzerne: caractérisation, zones de cultures, valeur culturale. Ann. Amélior. Plantes. 1, 210-255.
Pedersen, M.W. and Bohart G.E., 1953. Factors responsible for the attractiveness of various clones of alfalfa to pollen collecting bumble-bee. Agron. J. 45, 548-551.
Tasei, J.N., Picard, M. and Carre, S., 1978. Les insectes pollinisateurs de la luzerne (*Medicago sativa* L.) en France. Apidologie 9(3), 175-194.
Vansell, G.M. and Todd, F., 1946. Alfalfa tripping by insects. J. Amer. Soc. Agron. 38, 470-487.
Varga, P. et al., 1973. Lucerna - Cérès, Bucarest 302 p.

DISCUSSION

M.H. Poulsen *(Denmark)* You said that you were perhaps not in favour of the bulk, or even the population method of maintenance of genes and populations. Is your reservation based on the fact that the genes are not really available from these, or is your reservation because you think we are losing genes?

P. Guy *(France)* Bulk is a cheap way to keep an important genetic availability. However, the shift is continual and important. It is better to put together closed ecotypes. Bulk is a method of plant breeding to decrease crossing over but it is completely ineffective in keeping traits of adaptation or particular characteristics such as tolerance, earliness, reaction to injuries, day length, etc. We have to build as many bulks as desired forms.

Effect of temperature and moisture in the seed

S.ANDERSEN
University of Agriculture, Copenhagen, Denmark

INTRODUCTION

It is well known that the longevity of seed samples depends on temperature in the storage room and moisture in the seeds. The literature on this subject is tremendous, but reviews were given in three papers by Roberts (1973) and in a booklet by Justice and Bass (1978). In this short paper some results will be reported from long term storage at the Danish Agricultural University.

CROSS-POLLINATED SPECIES AT TWO DIFFERENT TEMPERATURES

Seeds of 50 commercial varieties of cross-pollinated species were delivered by plant breeders in 1965, and of each variety one sample was stored in an ordinary storage room with an average temperature of 12°C. Another sample was stored at -25°C. Before storage the samples were dried to a moisture content of 6 - 7%. Each year the germinating capacity was investigated in one sample of 100 seeds. When the germinating capacity in a sample in the cold store was below 70%, this sample was multiplied in an isolation plot in the field.

It has been necessary to multiply one sugar beet, three varieties of clover and one variety of grass, but samples of 45 varieties stored at -25°C are still germinating with a capacity of more than 70%.

On average, samples of *Beta* at 12°C fell below 70% germination after 11 years of storage, and samples of other species after 6 to 7 years (Table 2). The average germination capacity was 86% in those samples of *Beta* which were stored at -25°C and not multiplied. It was 92% for *Brassica*, 88% for legumes, and 90% for grass species. Eighteen seed samples stored at -25°C from 1965 to 1980 had a germinating capacity of more than 90% when tested in 1980.

TABLE 1

NUMBER OF SEED SAMPLES STORED FOR 15 YEARS AT TWO TEMPERATURES

Genus	Total	Samples multiplied	Samples with more than 70% germinating capacity after 15 years of storage 12°C	-25°C
Beta	13	1	0	12
Brassica	6	0	0	6
Trifolium and *Medicago*	12	3	0	9
Grass species	19	1	0	18

TABLE 2

LENGTH OF STORAGE PERIOD AT 12°C AND GERMINATING CAPACITY OF SAMPLES STORED AT -25°C

	Beta	*Brassica*	Legumes	Grasses
Average number of years with germinating ability above 70%. 12°C	11	7	7	6
% germination after 15 years storage at -25°C. Average of samples not multiplied	86	92	88	90

TABLE 3

NUMBER OF SAMPLES OF BARLEY, OATS AND WHEAT AFTER HARVEST IN DIFFERENT YEARS AND STORED WITH APPROXIMATELY 12% MOISTURE AFTER AIR-DRYING AND APPROXIMATELY 7% AFTER DRYING WITH HEATED AIR

	Dead samples				90% germination or more			
	Harvest year		Methods of drying		Harvest year		Methods of drying	
	1961	1962	Air	Heat	1961	1962	Air	Heat
Total number of samples in each group	81	162	81	162	81	162	81	162
After 5 years	0	7	7	0	57	88	16	129
" 10 "	14	42	56	1	3	65	0	68
" 15 "	60	58	37	81	0	20	0	20
" 18 "	79	73	71	81	0	3	0	3

CEREALS AT TWO MOISTURE LEVELS

Samples of wheat, barley and oats were harvested in 1961 and 1962. Twenty-seven samples of each species or a total of 81 samples were harvested 1961. The following year 162 samples were harvested and stored. Twenty-seven samples from 1961 and 54 samples from 1963 were air-dried to a moisture content of 11 - 12%. One hundred and sixty-two samples were dried by heated air to approximately 7% moisture content. A report of the methods used for harvesting and drying is given by Andersen and Andersen (1972).

After 5 years 7 air-dried samples had lost their germinating ability completely, but 129 samples out of 162 heat-dried samples had a germinating ability of more than 90%. After 10 years all air-dried samples germinated with less than 90%, but 68 of the heat-dried samples germinated with 90% or more.

The moisture content was very important for the longevity of the seed, but the year of harvest was important too. Seeds harvested in 1962 kept their germinating ability for more years than seeds harvested in 1961. The reason for this is not known, but the results indicate a great influence of weather conditions.

REFERENCES

Andersen, S. and Andersen, K., 1972. Longevity of seeds of cereals and flax. Acta Agric. Scand., 22, 3-10.

Justice, O.L. and Bass, L.N., 1978. Principles and practice of seed storage. Agriculture Handbook no. 506. US Dept. Agriculture.

Roberts, E.H., 1973. Predicting the storage life of seeds. Seed Science and Technology, 1, 499-514.

Roberts, E.H., 1973. Loss of seed viability. Chromosomal and genetical aspects. Seed Science and Technology, 1, 515-528.

Roberts, E.H., 1973. Loss of viability. Ultrastructural and physical aspects. Seed Science and Technology, 1, 529-545.

DISCUSSION

Luciana Quagliotti *(Italy)* Did you measure the mean germination time at the beginning of the storage period because this could give a good estimation of the vigour of the seed sample?

S. Andersen *(Denmark)* The germination capacity was estimated each year. We did not measure the mean germination time but it is quite clear that it will be prolonged. In my opinion, it is only a short time that you have a germination capacity above 80%, or even 75%. When it drops below this the seed will germinate a little more slowly than new seeds.

M.H. Poulsen *(Denmark)* This is a very interesting question. Do you still have the counts of these germination tests? Did you do them on two different days, because if so, you may be able to find some very interesting data to correlate with the longevity of the germination capacity?

S. Andersen Why should I have done it on two different days?

M.H. Poulsen Well, there is no obligation to do that but the testing authorities for germination use two days and I thought you might have done the same.

S. Andersen You are thinking about the method used at the Danish State Seed Control - no, we have not done that. I agree that it could probably give some information but our problem is that we are doing these germination tests ourselves and we do not have a room kept at a constant temperature. Therefore, there could be a variability in comparing results.

J. Jakubiec *(Poland)* I like this very quick method for drying seeds: you say 42°C, 24 h. Can this be used for any seeds - it does seem very quick?

S. Andersen I don't know what would happen if we used this method to dry grass seed. This was done with ears. These samples keep their longevity as long as samples treated in other ways.

B.H. Somaroo *(Syria)* How much information is available on the genetic deterioration in this long term storage of seeds?

S. Andersen Well, I think we must believe Roberts who has stated that genetic deterioration is proportional to the decrease in germinating ability. We do not have much information about this but he claims that it is so. The conclusion from this is that we should not wait until we get below 70% germinating ability. In the future we must try to multiply when we have lost only 10% of the germinating ability. Of course, it is not so simple as it sounds here because when you are doing a germination test it is most usual to use only 100 seeds and then your standard error, in some cases, gives you a different value from the true figure.

B.H. Somaroo Is it possible that there might be complications which you cannot detect in a germination test?

S. Andersen Yes, that is right, but again, as far as I know there is no one other than Roberts who has worked on this problem, at least on a large scale.

Luciana Quagliotti I would think that information on the genetic effect of seed storage could be obtained from one of the artificial methods of quick ageing of seeds.

J.T. Williams *(FAO)* I think it is a question of it being a relative test. When you experiment with actual years you have a time scale but the accelerated ageing test cannot be related to exact years.

Session 3
Technical aspects: Horticultural plants

Chairman: G.JENKINS

Genetic conservation by regeneration

HENNY ROELOFSEN
Institute for Horticultural Plant Breeding, Wageningen, Netherlands

Collection administration is a plant breeding activity. Both the collection administrator and the plant breeder work with genetic variation. They have different goals however. The plant breeder's object is to find and use the best from genetic variation, whereas the collection administrator aims at collecting and maintaining as much genetic variation as possible. This difference has fundamental implications for the working methods of plant breeders and collection administrators.

A plant breeder deals with the complexity he is confronted with by reduction: from the available material he selects only those plants for maintenance that meet his criteria. This results in a limited number of populations that are more or less homogeneous and that differ from one another in a restricted number of important characteristics.

However, the collection administrator is obliged to handle the whole of genetic variation, and it is one of his tasks to maintain this complexity.

Maintenance will generally include regeneration for multiplication or rejuvenation. <u>Multiplication</u> sometimes precedes long term storage and then it determines the potential genetic quality of a genebank accession. The maintenance of this potential genetic quality depends on the <u>rejuvenation</u> method that is to be used. Genetic quality may be thought of as the total number of different alleles in a population.

The purpose of both multiplication and rejuvenation is conservation of genetic variation by conserving the genes that carry this variation. It is impossible to regenerate a population with 100% certainty that no genes are lost. All a collection administrator can do is to restrict the inevitable loss of genes by using optimal regeneration methods for his material and his work conditions.

The collection administrator should adjust the regeneration method for certain material primarily to its breeding habit. The two extremes are complete self-fertilisation and complete cross-fertilisation. It is important to note that a collection administrator meets cross-fertilisation in more crops than a plant breeder does.

In the first place wild and primitive relatives of our cultivated crops may be outbreeding to a greater extent than the modern varieties and breeding lines on which the plant breeder focuses his attention. This applies to several important horticultural crops, for instance, tomato.

Second, populations of self-fertilising crops - such as beans - have a low percentage of cross-fertilisation. Since even a very low rate of cross-fertilisation is enough to cause heterogeneity at a considerable number of loci, it is necessary for genetic conservation purposes to treat such populations as genetically heterogeneous and cross-fertilising. The only populations that may be considered to be completely homogeneous and homozygous are lines of strictly self-fertilising populations - such as most lettuce populations probably are - and breeding lines and varieties obtained by repeated inbreeding. However, strict self-fertilisation is relatively rare, and the regeneration of breeding lines and modern varieties is not a main responsibility of the genetic conservationist. The following discussion of regeneration methods will therefore be based on the more complicated situation in genetically heterogeneous, cross-fertilising populations. They are assumed to fulfil Hardy-Weinberg conditions (random mating, no selection, no mutation, no migration), so that we may use population genetic rules to help us to judge alternative regeneration methods.

The basic phenomenon that we have to deal with when regenerating a small population is <u>genetic drift</u>. Genetic drift is the loss of genes due to random fluctuations of gene frequencies that occur in small populations. Loss of genes is directly connected to loss of heterozygotes, and this is the phenomenon used in quantifications. In a random mating population 1/2 N heterozygotes are lost per generation; N is the population size. For self-fertilisation N = 1, so that half of the heterozygotes are lost per generation. If a collection administrator is willing to accept 1 - 2% loss of heterozygosity per generation, he will need a population size of 30 - 50.

At IVT this is realised for insect pollinated crops such as cabbages. They are placed in netting isolation rooms or cages of different types, of sizes varying from 6 - 30 m^2.

The plants are isolated just before anthesis. Both flies and bees are used for pollination in order to approximate random mating as closely as possible by maximal cross-pollination. Flies are more independent of weather conditions, but bees only concentrate on the flowers when they are active. This method is relatively little labour consuming and simple. The same method and equipment may be used for carrot, onion and radish populations.

For a cross-fertilising population we should aim at using the same population size through the generations. Changing population sizes leads to an intermediate effective population size $N_{e(c)}$, which is always closest to the smallest actual value. $N_{e(c)}$ is the harmonic mean of the population sizes in the different generations:

$$\frac{1}{N_{e(c)}} = \frac{1}{t} \left(\frac{1}{N_1} + \frac{1}{N_2} + \ldots\ldots + \frac{1}{N_t} \right)$$

If the population sizes in three generations are for instance 100, 10 and 100, the effective number is 25. This means that the same loss of heterozygosity would have occurred if a population of 25 individuals was used in all three generations, which would thus have been more economic after all.

In the IVT situation an incidental population size of 10 would lead to an effective population size of 21 in three generations if the normal population size of 50 has been realised in the other two generations.

Another outbreeding crop of which IVT keeps a collection is spinach. There are two essential differences between this crop and those of the previous group. The first is its wind pollinating habit which requires a different isolation technique.

Multiplication has to be carried out in airtight plastic or glass rooms or cages with filters in the ventilation openings. To ensure maximal cross-pollination, a forced air circulation is needed. Lacking such equipment, IVT regenerates spinach populations by a combination of isolation by distance and isolation by obstacles. The populations are placed in separate, normal greenhouses, as far away from one another as possible.

The second point that we have to be aware of is that spinach may at least have partially separate sexes, whereas the crops of the previous group were all hermaphrodite. In a dioecious crop the number of males and the number of females are not necessarily the same. If they differ, the random fluctuations in the gene frequencies of the rarest sex are enlarged in the next generation. Therefore in spinach the loss of heterozygosity per generation is not directly related to the actual population size as it is in cabbages, but to the effective population size N_e.

N_e is the number of plants that gives rise to a loss of heterozygosity of $1/2\ N_e$. The effective population size of a population consisting of N_m males, N_f females and N_{mf} hermaphrodite individuals may be calculated from the following expression:

$$\frac{1}{N_e} = \frac{N_m (N_{mf} + N_f)^2 + N_{mf} (N_m + 2N_{mf} + N_f - 2)^2 + N_f (N_m + N_{mf})^2}{4((N_m + N_{mf})(N_{mf} + N_f) - N_{mf})^2}$$

In the case of completely separate sexes N_{mf} is zero and the formula becomes:

$$\frac{1}{N_e} = \frac{1}{4N_m} + \frac{1}{4N_f} \quad \text{or} \quad N_e = \frac{4N_m N_f}{N_m + N_f}$$

If there are as many males as females all individuals reproduce equally and the effective population size equals the actual population size. If the number of individuals of one of the sexes is smaller than of the other sex, this will lead to an effective population size that is smaller than the actual number of individuals. This is shown in Figure 1.

IVT normally uses 60 plants per population, which results in a probability of about 95% of an effective population size of more than 57 in the case of completely separated sexes.

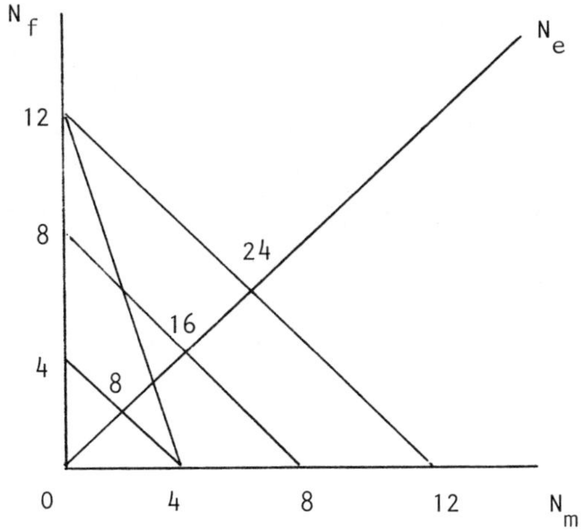

Fig. 1.

The examples given here concern crops that are indigenous in The Netherlands. They can be grown outside or in unheated glasshouses and their seed growing techniques are known and relatively cheap. The second group in which we deal with heterogeneous, outbreeding material varies in these respects. It consists of crops that were introduced from the tropics or the subtropics. At our latitude they are preferably grown in greenhouses, both for commercial production and for seed production. The most important examples of this group are tomato, cucumber, pepper and eggplant. The modern, advanced material is practically completely self-fertilising, but the wild and primitive material is cross-fertilising to a considerable extent and should therefore be treated in the way discussed above for cabbages. Regeneration of such material is frequently hampered by lack of suitable growing and isolation facilities.

A method that reduces the number of plants and thus the space required for multiplication is grouping. By grouping the plants of a population into smaller, randomly mating subgroups that are isolated from each other, the effective population size may exceed the actual population size. The general formula to be used for calculation of the effective population size is:

$$\frac{1}{N_{e(gr)}} = \frac{n^1 - 1}{kn^1 - 1} \times \frac{1}{n_e}$$

in which k is the number of groups, n^1 is the number of descendants of one group that will constitute the next generation and n_e is the effective population size of one single subgroup. $N_{e(gr)}$ is the effective population size of the whole subdivided population.

Table 1 gives $N_{e(gr)}/N$ for different values of k, n_e and n^1.

TABLE 1

	k = 2 n_e = 12	k = 4 n_e = 6	k = 6 n_e = 4	k = 8 n_e = 3	k = 12 n_e = 2
n^1 = n	1.05	1.15	1.28	1.44	1.92
n^1 = 2n	1.02	1.07	1.12	1.18	1.31

The more a population is split into smaller groups, the more effective is the grouping, especially when the population size is constant through different generations. The last column in Table 1 shows that crossing pairs is most effective.

Pair crossing may be executed by numbering the plants of a population and crossing plant 1 with plant 2, plant 3 with plant 4, and so on. The population size remains the same if two descendants per pair constitute the next generation (Table 2).

TABLE 2

PAIR CROSSING

generation t	1 x 2	3 x 4	....	n - 1 x n
generation t + 1	1 , 2 ,	3 , 4 ,	....	n - 1 , n

An equivalent method that can also be applied to an odd number of plants is chain crossing. For chain crossing a population of three plants, plant 1 is crossed with plant 2, plant 2 with plant 3 and plant 3 with plant 1. The population size is constant if the next generation is made up of 1 descendant of each cross (Table 3).

TABLE 3

CHAIN CROSSING

generation t	1 x 2	2 x 3	....	n x 1
generation t + 1	1 ,	2 ,		n

Both for pair crossing and chain crossing we may use the formula:

$$N_e = \frac{Ng - 2}{g - 1}$$

in which g is the number of gametes that one individual contributes to the next generation. When the population size is constant g equals 2, so that:

$$N_e = 2N - 2.$$

This results in the following values of N_e for different N's:

N :	3	4	5	6	7	8	16	26	32.
N_e:	4	6	8	10	12	14	30	50	62.

This relation between N and N_e is represented by a simple diagram (Figure 2).

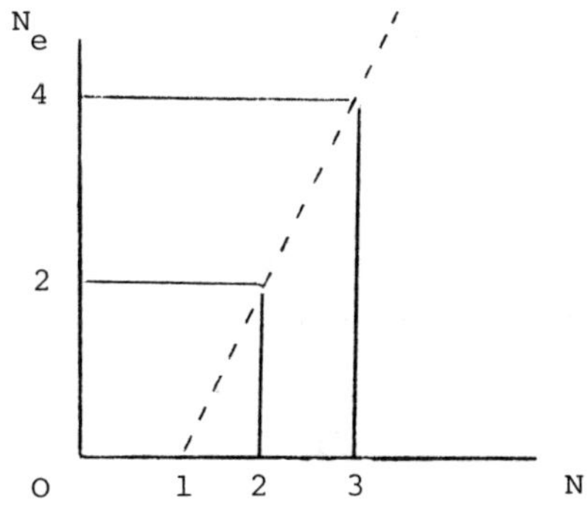

Fig. 2.

Although chain crossing and pair crossing with g = 2 lead to the same effective population size, chain crossing is to be preferred over pair crossing if cytoplasmatic factors are taken into account.

In order to restrict the loss of heterozygosity per generation to 1 - 2%,16 - 26 plants have to be chain crossed or pair crossed, compared to 30 - 50 when one random mating group is used. The price for this reduction has to be paid in labour. Chain pollinations are carried out by hand and the pollinated flowers may have to be bagged. The pair crossings have be isolated with insects or pollinated by hand, if necessary followed by bagging. However, the most laborious part is the administration, because the progeny from different pairs have to be kept separated. And then of course some extra sowing and planting has to be done as well in order to be sure that one plant of each parent or two plants of every pair reach the reproductive stage. This is necessary, because it is very important not to change the population size during the generations as we noted above, and in this system every accidental change means a definite loss.

Besides the relatively large labour demand, this method causes a special problem at our Institute; the required number of plants does not fit into the common growing and seed growing practice. Sixteen to twenty-six plants is still a multiple of the normally used 5 plants, leading to an effective number of 8 when chain crossed.

In the existing system genetic conservation is only possible if the original, heterogeneous population is split into at least 4 sub-populations that are maintained separately by chain crossing at least 5 plants per generation. It is logical and practical to take different morphotypes as sub-populations if possible. This will help the breeder to use the available genetic variation as efficiently as possible.

Practical conclusions that may be drawn from the above are:

1. For genetic conservation purposes populations should be treated as genetically heterogeneous and cross

fertilising, except those which are known to be strictly self-fertilising.

2. An effective population size of 30 - 50 leads per generation to a loss of heterozygosity of 1 - 2%.
3. The effective population size exceeds the actual population size if grouping, pair crossing or chain crossing is executed.
4. It is most efficient to use the same effective population size in each generation.
5. The regeneration administration should record both the number of plants and the pollination system used, so that the degree of maintenance of the genetic variation of a population can be established.
6. The number of plants required for regeneration of cross-fertilising populations for population genetic reasons generally exceeds the number required for the production of the seed quantity IBPGR prescribes for long term storage.
7. Genetic conservation is most effective when collection and regeneration methods correspond.
8. The contents of a collection are not fully represented by the number of its accessions; their genetic quality is at least equally important.

REFERENCES

Anonymous, 1976. Report of IBPGR working group on engineering, design and cost aspects of long term seed storage facilities. International Board for Plant Genetic Resources, Rome.

Banga, O., 1953. Inleiding tot de plantenveredeling. Tjeenk Willink, Zwolle.

Crow, J.F. and Kimura, M., 1970. An introduction to population genetics theory. Harper and Row, Publishers, New York.

Drijfhout, E., 1981. Maintenance breeding of beans. Unpublished.

Hackbarth, J., 1961. Spontane Frembestäubung bei Selbstfefruchtern. Vortrage für Pflanzenzüchter. DLG-Verlag, Frankfurt/M.

Stam, P., 1981. Personal communication.

Van der Veen, J.H., 1980. Populatiegenetica (Theoretische grondslagen).

Zeven, A.C. and Van Harten, A.M., 1979. Broadening the genetic base of crops. Pudoc, Wageningen.

DISCUSSION

Y. Demarly *(France)* With regard to the domestication of sub-groups in a population, I think if you want to preserve the integrity of a heterogeneous cultivar or population the elements of the sub-groups must be a random sample of each.

Henny Roelofsen *(Netherlands)* That would make the work more difficult for the breeder.

Y. Demarly Yes, but if you do not have a representative sample of the population, the next generation will change, even if you put together the progeny of each sub-group. There are two alternatives: if you choose integrity and an exact representation of the initial cultivar, you must have a representative sample in each sub-group. If you choose to preserve a large genetic diversity, you have to choose your sub-group from the initial population. You must increase the genetic distance between each sub-group so you have to characterise phenotypically some types in the population and each of your sub-groups must be typical. At each generation you have to increase the genetic distance between each sub-group. Then, when you cross the progeny of each separate sub-group, you will increase the genetic distance between each sub-group. Then, when you cross the progeny of each separate sub-group, you will increase the genetic diversity each time. You have to choose between the methods.

Henny Roelofsen I think it is very interesting, yes.

E.L. Breese *(UK)* This part of the discussion is of fundamental importance in looking at the way in which we can preserve the populations. At Aberystwyth we have elected the random pollination of a random sample. If I understand him correctly, Professor Demarly is advocating an alternative way of assortative mating which would indeed increase the variance of the population as we go along. I am slightly worried that that might also destroy the balance and the structure of any population to some extent. I would be particularly worried about grasses and the perennial groups.

Y. Demarly I agree that in this way we will destroy the structure and you will have a kind of disruptive selection.

Henny Roelofsen But I think it is not possible to keep the original structure just as it was, so if you are going to influence it anyway, you might as well do it in the way that is most useful to the breeder.

G. Jenkins *(UK)* I wonder if anyone would wish to dispute that proposition. Do we have genetic conservationists here who do not regard themselves as breeders?

J.G. Hawkes *(UK)* I fear I am jumping into a subject that I am not too well qualified to comment on, but the basis of our conservation strategy has always been considered as trying to preserve, as far as possible, the structure of the populations that are put into storage and which have to be regenerated from time to time. If this can be done, and we realise that a certain loss of heterozygosity must occur and losses of alleles, then it seems to me that we should try as far as possible to retain population structure as it was when it went into store. I am sure Dr. Roelofsen would agree with that. What we can do for the breeders is fine, but what slightly worries me is that breeders have slightly different ideas and different objectives from human generation to human generation. Twenty years back they may have wanted certain things, and twenty years on, certain other things. Other perspectives, other objectives, come forward, naturally enough. Therefore, perhaps we are better advised to try to keep our populations as they were rather than to try to make them available for breeding objectives which apply at this point in time but which are not necessarily the objectives that will apply very much later. Perhaps what you are saying, Dr. Roelofsen, does not conflict with that. I would like to hear your comments.

Henny Roelofsen You mean that we have to protect the breeders against themselves in a way?

J.G. Hawkes I would not have put it quite so strongly.

G. Jenkins Perhaps it is more a case of protecting the populations against the breeders!

J.G. Hawkes There are differences in what is consistently correct as breeding objectives at any one time. They vary, naturally.

Henny Roelofsen I think the breeder will always start from morphological characteristics in the first place. It is logical to make morphological sub-groups and that does not conflict with genetic conservation so why not do that? It makes it easier to control mistakes, mixing of populations and so on. The breeder may change but he will always want morphological, homogeneous populations. So, if we have to split the population anyway, I propose that we do that and conserve the other alleles within that population as well as we can.

J.G. Hawkes Forgive me if I am saying something out of turn but I always assumed that the breeder was not too interested in morphological characters, that he was looking for characters of physiological adaptation and resistance to diseases and pests, which don't show up in the morphotypes in any case. So we have to keep that in mind surely?

Henny Roelofsen Yes, we must keep that in mind, that is what the breeder is looking for but only if the other characters are there already. If he doesn't have morphological homogeneity, we will have to start from that. Most breeders do have the luxury of homogeneous populations that they can start from, and we should keep it that way I think.

J.G. Hawkes We'll talk about this later.

J. Jakubiec *(Poland)* You mentioned the method of pollination, gathering the pollen from the older plants and pollinating each of the plants belonging to these populations with the mixing pollen. Such a method is very good if you have fully incompatible plants and therefore it is a much better method that you propose to make a pair, 1 to 2, 2 to 3, 3 to 4, and so on. In this case it is better to use the random pair.

Henny Roelofsen Yes, I think you are right.

W. Plarre *(Federal Republic of Germany)* I would like to make a comment on the question of the morphological and physiological characters in which the breeders are interested. I don't think we should overestimate this formalism in breeding. For the future I think we should concentrate more on a homogeneous type in which we have uniform biochemical characters. We know from the regulations for releasing cultivars that we should have a homogeneous type but I would say that for storage and for maintaining, it is not so important to have morphological uniform characters in the future but rather more uniform physiological and biochemical characters.

Henny Roelofsen I think morphological homogeneity will always stay as a prerequisite for varieties. Biochemical homogeneity will come on top of that.

W. Plarre No, I think we should overcome this old idea of formalism.

Henny Roelofsen Maybe you are right, but I think you are asking too much. For practical reasons as well it is handy to have a homogeneous lot to multiply and to handle as a unit.

E.L. Breese Sometimes we tend to get projected in an area of conflict where there is no real conflict. Many of us are here as breeders and as conservers and in one sense we appear to wear two hats, but we don't. As a breeder, I would always like to preserve a population in its original entity so that in future I, or other people, could take a decision on what I wanted to do with it. In that sense I reaffirm what Professor Hawkes put forward, that really our aims should be to preserve the population as it is. In that sense I do not go along with methods of sub-grouping which would break up the population.

Henny Roelofsen But in your case, for example, if you use 25 plants, you are going to lose alleles anyway, and you have to be aware of that.

E.L. Breese As I tried to point out yesterday, we are trying to preserve gene assemblies which have their alleles maximised for certain expression.

Henny Roelofsen But you can never estimate all alleles.

E.L. Breese No, but the majority of the ones that are important for that particular adaptation that we are seeking to conserve. Very often we are looking at gene frequencies amongst the whole mass of polygenes or gene sets which are way above 10%. If we can preserve most of those, that is the genetic resource the breeder is going to be looking for.

Henny Roelofsen But that is not the original population that you were working with; it is a selection from that population.

E.L. Breese We have talked about physiological characters, morphological characters, biochemical characters. As always, now and in the future, the problem for the breeder is to decide what his criteria are going to be. That is always problematical in the improvement in crops, particularly advanced breeding systems. What I am saying is that you cannot recognise, phenotypically, what the population contains. Again this reaffirms the need somehow to freeze the population genetically.

Principles of genetic sampling and conservation in perennial species A case study in arctic bramble (*Rubus arcticus* L.)

J.TAMMISOLA
University of Helsinki, Finland

ABSTRACT

Principles of genetic sampling and conservation in perennial species are considered and, as an application, arctic bramble (Rubus arcticus L.) is discussed. This is a subarctic wild berry species now in the process of domestication. Possible ways of conserving its genetic variation in the Nordic Genebank are considered.

Vegetative propagation of clones, both naturally and with the aid of mist techniques, is common practice with the arctic bramble although it is rather expensive. In the near future arctic bramble clones may be stored as meristem cultures under refrigeration, and later on may also be deep frozen.

Intercrossing of arctic bramble clones by hand is too resource-consuming and does not offer essential advantages taking into consideration the total strategy of conservation.

Selfing is practically ruled out in the arctic bramble, due to the strong self-incompatibility and inbreeding depression. Instead, bulk-pollinating will probably prove to be both easy and cheap and will still maintain the most fundamental genetic structures.

Thus a dual strategy is proposed for the arctic bramble - both clone collections and bulk populations should be established.

1. *Clone collections should be established during the first stage.*
2. *As a basis for classifying the collection localities in the northern countries, a subdivision into five natural geographic zones, characterised by vegetation and soil types, should be used.*
3. *It is anticipated that these zones will influence the formation of genetic co-adaptation.*
4. *In clone collections of R. arcticus, accessions from a particular zone should be planted together to secure the conservation of co-adapted genes.*
5. *Different zones in a clone collection should be isolated by a surrounding zone of a triploid clone which flowers abundantly but is incapable of fertilisation, either as a male or as a female parent.*

INTRODUCTION

Collection, storage and regeneration are inter-dependent in many ways. Therefore, we should keep in mind the general plan when designing the details. Thus, when considering the alternatives, for example, in the regeneration phase, we should formulate the total strategy of genetic conservation in the species or species group. The overall picture must also be considered since species are extremely complicated systems, still far from being well understood genetically.

The germplasm reservoir of a species and the total strategy of its conservation depend largely on its evolutionary background, its ecology and population biology, as well as on the history of its cultivation.

Attention should be given to the following points:

What is the total amount of genetic variation in the species or the species group? More accessions warrant collection and maintenance in a more variable species. On the other hand, relatively more labour is required in a less variable species, in order to recover enough of its genetic variation for conservation.

What is the genetic structure of the populations in the species, whether they be natural or cultivated? This affects both the collection plans (number and choice of populations, sample size and sampling method) and the methods of maintenance (should bulk populations, more narrowly based populations, or individual genotypes, be used?).

What is the mating system? Is the species self- or cross-pollinated, is the pollen transferred by wind or insects, does some degree of incompatibility occur? Population structure is influenced by the degree of natural self-pollination, and so is the amount of inbreeding depression following artificial self-pollination. Wild species and primitive forms are often self-incompatible. In such cases it is impracticable to maintain alleles via forced self-pollination, because obligate outbreeders commonly suffer from severe inbreeding depression. Isolation requirements depend on the mating system as well as on the distribution of wind velocity and direction, or on the foraging habits of pollinating insects.

In the first place, asexual means of reproduction influence the genetic structures of populations and secondly, they also make maintenance of individual genotypes possible in cross-pollinated species.

How long is the generation interval of the species? In perennials it is sometimes very long which makes storage of breeding material as seeds impracticable. Artificial shortening of the generation interval has proved possible in some cases.

What is the degree of domestication in the species? Species which have been cultivated for a long time often have a large number of narrow, thoroughly explored, genetic units to be preserved (varieties, special genotypes), whereas freshly domesticated species usually have only a few.

What is the genetic structure of the cultivar of the species? If the cultivars are populations, then population-based methods of conservation are essential, e.g. bulk methods. If, on the other hand, the cultivar consists of a single genotype, methods involving the maintenance of individual genotypes will also be required.

What is the object of conservation? If we are interested in the preservation only of alleles of certain loci, then bulk populations will suffice. If, in addition, we wish to retain higher level genetic structures (co-adapted gene complexes, genotypes), then genetically more narrowly based populations, even individual genotypes, should be regenerated. This question also has a bearing on the choice of a suitable storage technique; genotypes cannot be maintained unaltered via seeds except in strictly self-pollinated or apomictic species.

What techniques of collection and storage will be possible in the species, at least in the near future? Collection and long-term storage is easiest by means of seeds - if seeds are produced and keep their germinability long enough. Vegetative parts may sometimes be cold stored, but at most for a few years. Long term-storage as tissue cultures (usually meristems) has proved possible in many plant species. However, a considerable amount of labour may be required to develop adequate methods in a new species. Maintenance through continuous cultivation, either by sexual or asexual means, is possible but has several disadvantages. Nonetheless, in the case of perennial woody species, e.g. *Ribes* sp., it may be necessary to establish clone banks.

The advantages and disadvantages of the various techniques of collection and storage can be assessed in the following ways:

1. How much of the genetic variation can be collected and maintained per unit of available resources?
2. How great is the risk of genetic losses?
3. What is the longevity of propagules, i.e. how long is the regeneration interval of stored stocks?
4. How are the allele frequencies in the populations, the structure of the co-adapted gene complexes and the genetic constitution of the individual genotypes, affected by environmental conditions during regeneration?
5. What is the adequate size of a sample, and how many populations ought to be conserved?
6. What is the risk of contamination?
7. How can diseases and pests be controlled?
8. What kind of observations for the evaluation of the accessions can be incorporated into the technique in question?

THE ARCTIC BRAMBLE *(Rubus arcticus* L.*)*

The total extent of genetic variation in the arctic bramble

The arctic bramble *(R. arcticus* L.*)* belongs to the sub-genus *Cylactis* (Focke) of the genus *Rubus*. According to Focke (1910), *Cylactis* comprises 14 species, of which three (*R. arcticus* L., *R. saxatilis* L. and *R.humulifolius* C.A. Meyer) occur in Finland. The closest relatives of the arctic bramble, *R. stellatus* Sm. and *R. acaulis* Michx., are nowadays regarded as subspecies of *R. arcticus* (Hulten, 1968; Larsson, 1969; Kallio, 1975; Kotimaki and Hirsalmi, 1979.

The arctic bramble is circumpolar in distribution. Ssp. *arcticus* is found throughout subarctic Eurasia, and the distribution extends to the northern parts of North America, into Alaska. Ssp. *acaulis* occurs in the north eastern-most tip of Asia from the shores of the Bering Straits, across Alaska and into sub-Arctic Canada. Ssp. *stellatus* has a narrow distribution range in north western Alaska, the Aleutian Islands and into north eastern Asia on the shores of the Bering Straits, roughly where the distributions of ssp. *arcticus* and ssp. *acaulis* coincide (Figure 1).

The subspecies *stellatus* and *acaulis* differ from *arcticus* and from each other with regard to many characters. For example, the characteristic aroma of arctic bramble is practically absent in both *stellatus* (Kallio, 1975) and *acaulis* (Larsson, 1955), whilst the growth habit of *stellatus* is considerably more 'weedy' than in *arcticus*.

Polyploid forms of the species, mainly triploids and tetraploids have been induced or found occurring naturally, as well as hybrid forms involving other species, sometimes belonging to other subgenera (Larsson, 1969; Vaarama, 1949).

On the basis of preliminary results (Tammisola, unpublished) *R. arcticus* appears to show considerable variation within Finland.

Germplasm collections

At present there are about 10 to 20 arctic bramble strains in Sweden collected and worth conserving, a few polyploid forms and a large number of species crosses. In addition, a single clone of a *stellatus* type and two varieties derived from the cross *stellatus* x *arcticus* are maintained in Sweden (Larsson, personal communication). In Finland, 12 or more of A. Ryynänen's strains are being maintained, 3 of which have achieved varietal status. Tammisola has about 150 distinct arctic bramble genotypes from an originally collected total of 1200 samples (Figure 2) which included many duplicates. Tammisola's collection also includes one triploid and several *saxatilis* hybrids. At Piikkiö there are a number of hybrids with *R. idaeus* and a single variety derived from the cross *stellatus* x *arcticus*. At the Komarov institute in Leningrad there is reputed to be an especially interesting self-fertile strain of the arctic bramble (Hardh, 1976).

Most of the genetic variation of *R. arcticus* - comprising East European, Asian and North American forms of ssp. *arcticus* as well as ssp. *acaulis* in its entirety - has yet to be introduced to germplasm collections. Ssp. *stellatus* is probably represented in collections by a single clone, which has been extensively used in breeding programmes.

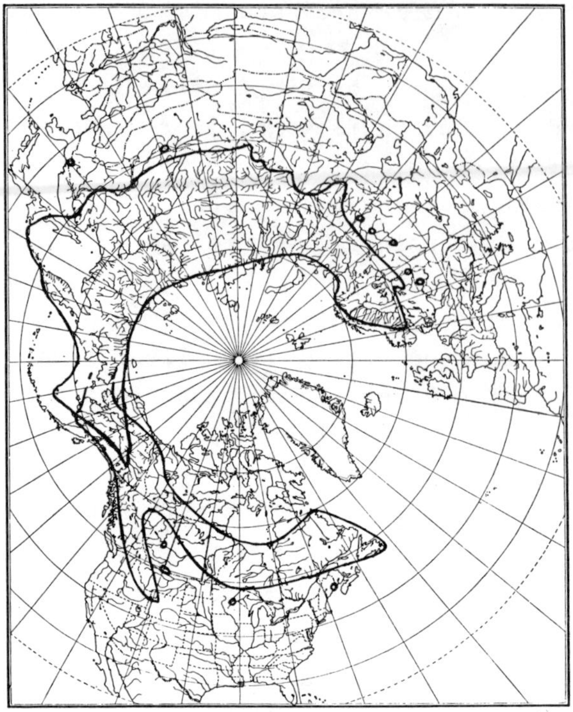

Fig. 1a

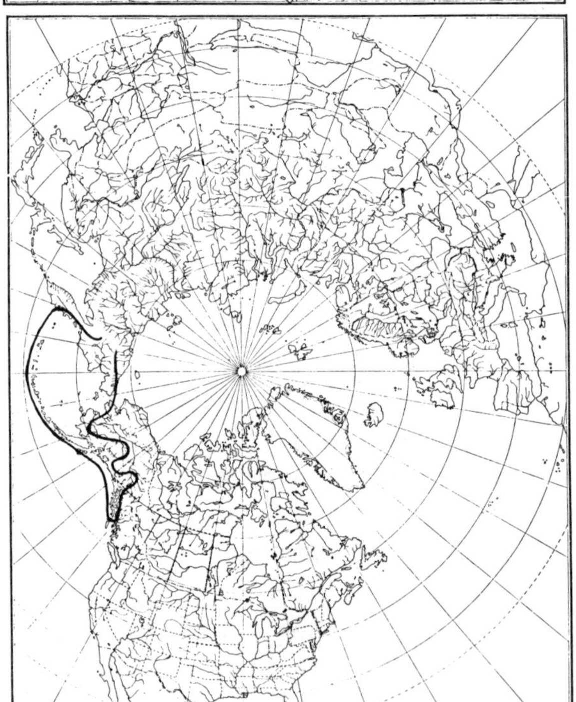

Fig. 1b

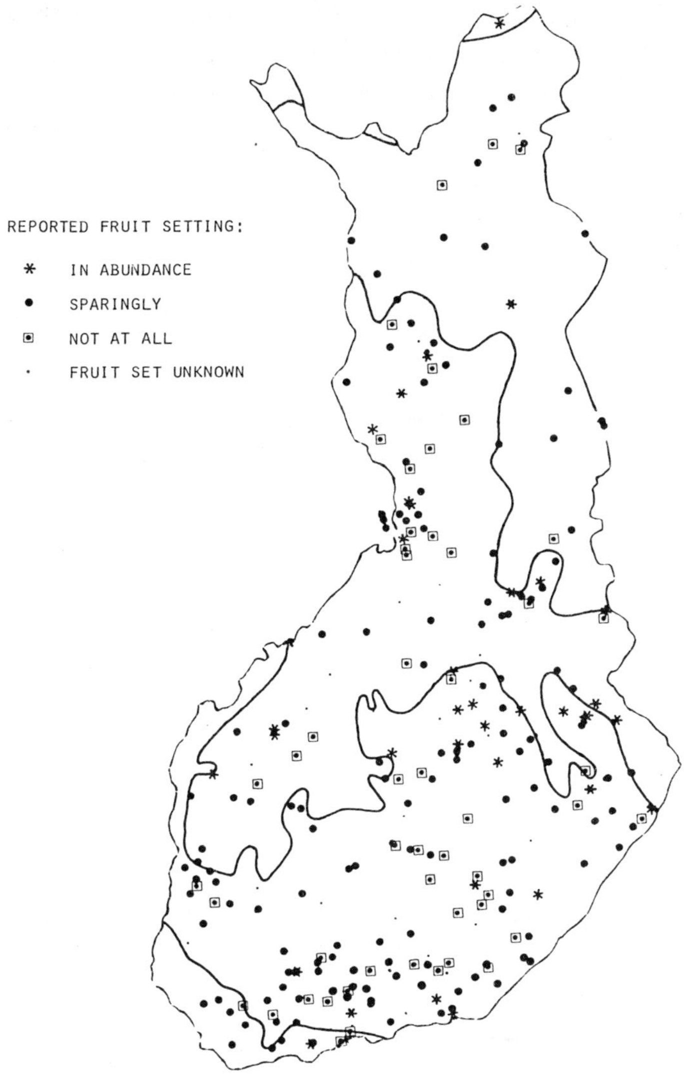

Fig. 2. A survey of *R. arcticus* populations in Finland in 1976

The genetic structure of populations

Preliminary results (Tammisola, unpublished) indicate that arctic bramble populations often consist of a single clone. In such populations, berries will not be produced because the

---

Fig. 1. General distribution of *R. arcticus* and its subspecies (Hulten, 1971, modified).

arctic bramble is a self-incompatible species. Even the berry-bearing populations generally appear to consist of few genotypes (Figure 3). As the population ages, the genetic structure becomes even more simple as the vegetatively most vigorous clone or clones take over.

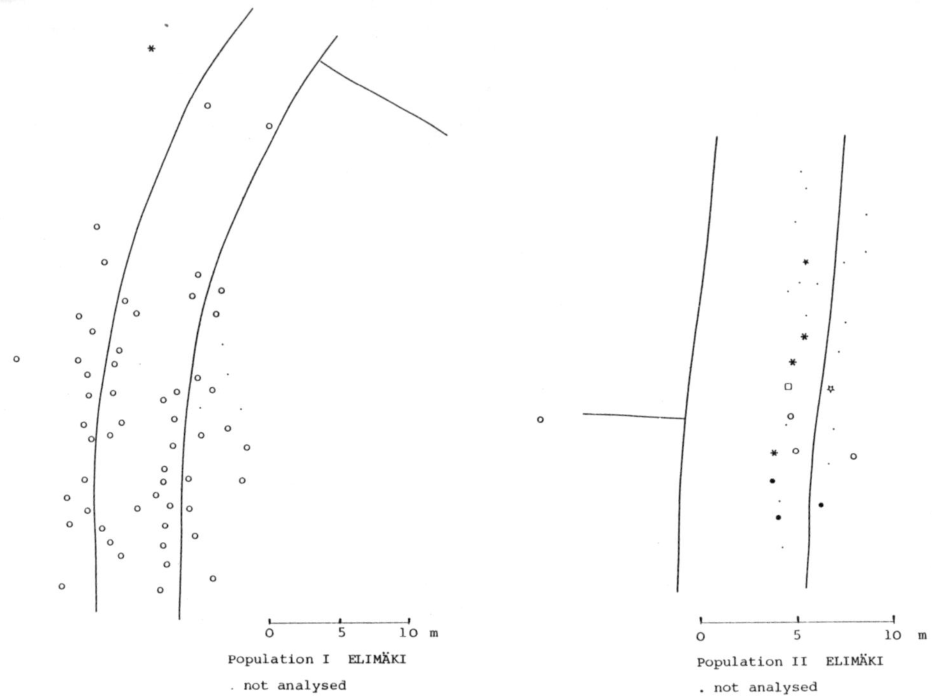

Fig. 3. The clonal structure of two arctic bramble populations, situated 600 m apart along the same forest road. Population II set fruit abundantly, Population I not at all.

The genetic content of arctic bramble populations is thus narrow. Notwithstanding this fact, the powerful self-incompatibility mechanism and severe inbreeding depression in this species indicate that naturally occurring clones are probably not very inbred. The former extent of variation is totally unknown, likewise how much of it is still present in the form of seeds possibly remaining dormant for decades. In the past, considerable migration of seeds through the agency of birds was quite likely. However, as berry production has fallen, the migration of seeds from one site to another in this way is probably of little significance at the present time. According to some observations, birds do not appear to be interested in berries in areas where few are produced.

When collections are made, it would appear to be worthwhile to take only a small sample of each population, and one should try to sample as many populations as possible. Likewise, if one is considering genetic conservation of material as populations, one should use a considerably wider genetic base to form the conserved population rather than form it on the basis of a single occurrence.

The mating system

The arctic bramble is a strictly cross-pollinating, strongly self-incompatible species with a gametophytic one-locus system of incompatibility. This minimises inbreeding in natural populations, although since the population consists of only a few genotypes, inbreeding cannot be totally avoided. Inbreeding depression is fairly severe (Ryynänen, unpublished), therefore artificial self-population of a strain (by circumventing incompatibility) is not a suitable means of conservation of separate alleles. In practice, by far the most important pollinators are bumble bees and honey bees (Warming, 1886; Poppius, 1903; Silen, 1906; Saastamoinen, 1930; Ryynänen, 1973).

Asexual means of reproduction

The arctic bramble spreads vegetatively by underground stolons, at the rate of a few decimetres a year (Saastamoinen, 1930). Even large populations can theoretically be composed of a few, even a single, genotype. In diploid *Rubus* species such as *R. arcticus* (2n = 14), apomixis does not occur (Knight and Keep, 1958).

The generation interval

Without the aid of a greenhouse, the time needed from seed to seed is about two years for the arctic bramble (Ryynänen, 1973, or as much as 4 - 5 years, depending on conditions of cultivation (Larsson, 1969). To obtain a sufficient stand size for evaluation (0.5 - 1.0 m^2) takes about 4 years (Ryynänen,

1973). It should be possible to reduce these times somewhat by using a greenhouse (in the initial stages). A limitation here is the obligatory period of winter dormancy required by this species. Also, during the winter, greenhouse conditions are conducive to outbreaks of disease.

The degree of domestication

The arctic bramble is in the initial stages of domestication. There are only a few varieties and these are relatively primitive: selected natural strains of ssp. *arcticus*, their primary crosses and the hybrid progeny from ssp. *stellatus* x ssp. *arcticus* crosses. There are not many special genotypic forms - mainly polyploids and a few species crosses. Methods of cultivation and utilisation are in such a state of flux that any definition of desirable agronomic characters is difficult (Larsson, 1955; Ryynänen, 1972, 1973; Hiirsalmi, 1974).

The cultivar

In the arctic bramble, the variety may comprise a single genotype due to self-incompatibility but two varieties must be cultivated side by side. In breeding work, the entire broad sense heritability is exploitable.

In theory, it should be possible to use a seed-raised population as a variety, provided the variety conforms with the uniformity demands of varietal legislation. This would at least solve pollination problems: the planting system involving two varieties has proved to be uncertain with regard to pollination because one or other of the varieties tends to be poorly adapted to local or seasonal conditions. Such problems seem to have occurred locally when the first pair of varieties was cultivated in Finland. These varieties were Mespi (from Piikkiö on the coast of south west Finland) and Mesma (from Maaninka, near Kuopio, 300 km farther north which is in a different natural geographic zone where a more continental climate prevails (Figure 4). A population-based variety would possess a considerably higher degree of adaptation to various conditions, and continued natural selection would ensure that local adaptation would progressively increase with each succeeding generation.

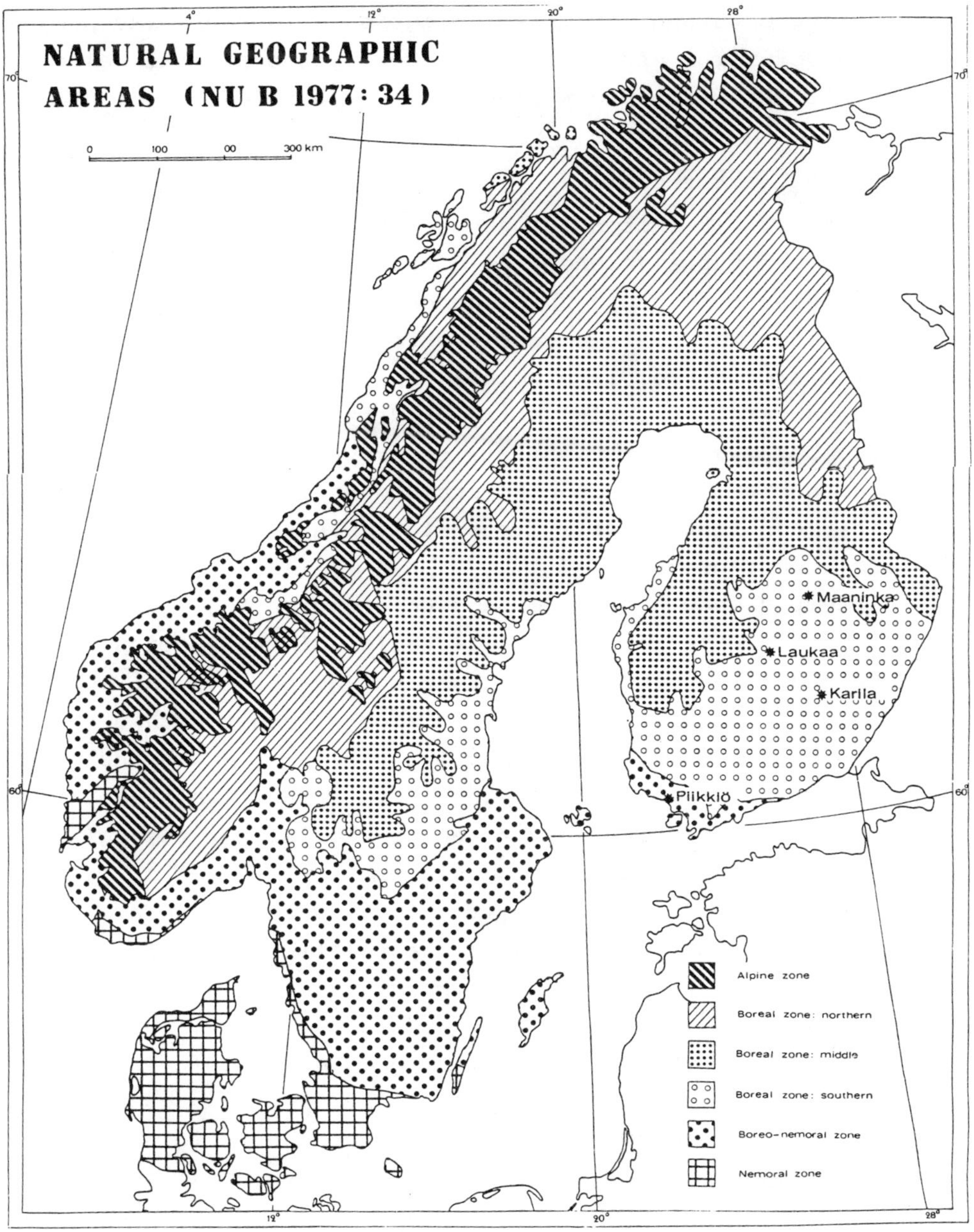

Fig. 4. The natural geographic division of zones within the northern countries (Anon, 1977)

The raising of stands from seed might slow down or even prevent the onset of virus and fungal diseases due to natural selection where infected individuals produce fewer berries.

The object of conservation

In the first instance, one should collect and conserve as many wild genotypes (clones) as possible, and evaluate them in field trials. From among these genotypes it may be possible to select clones which are directly suitable for cultivation, since the arctic bramble displays considerable variation for many characters, including berry production. In contrast to the blackberry in which apomixis is prevalent, in the case of the arctic bramble one cannot retain genotypes in seed form as genetic recombination will immediately break up combinations of desirable characters.

Attempts should also be made to conserve the most fundamental part of the genetic co-adaptation of the species. To provide the framework for preserving co-adaptation, a recently published natural geographical division of zones within the northern countries could be used (Figure 4) in which the occurrence of arctic bramble falls into five areas. As these areas, or zones, are designated according to floristic type, soil type and climate, such a division should provide a means of conserving genetic co-adaptation (Allard, 1975) within each area. In practice, one would compose a bulk population to include genotypes from a given natural geographic area. This would give five bulks in all.

In order to pursue this aim, clones maintained in a collection would be planted in the field in five groups corresponding to their natural geographic area of origin (Figure 5). In order to prevent cross-pollination between those originating from different areas, the groups would have to be effectively isolated from each other. The basic seed for each bulk could then be harvested directly from each group of clones.

In order to prevent the disruption of co-adaptive gene complexes, each bulk population must subsequently be maintained under the conditions prevailing within its own geographical area, i.e. at some locality within the area.

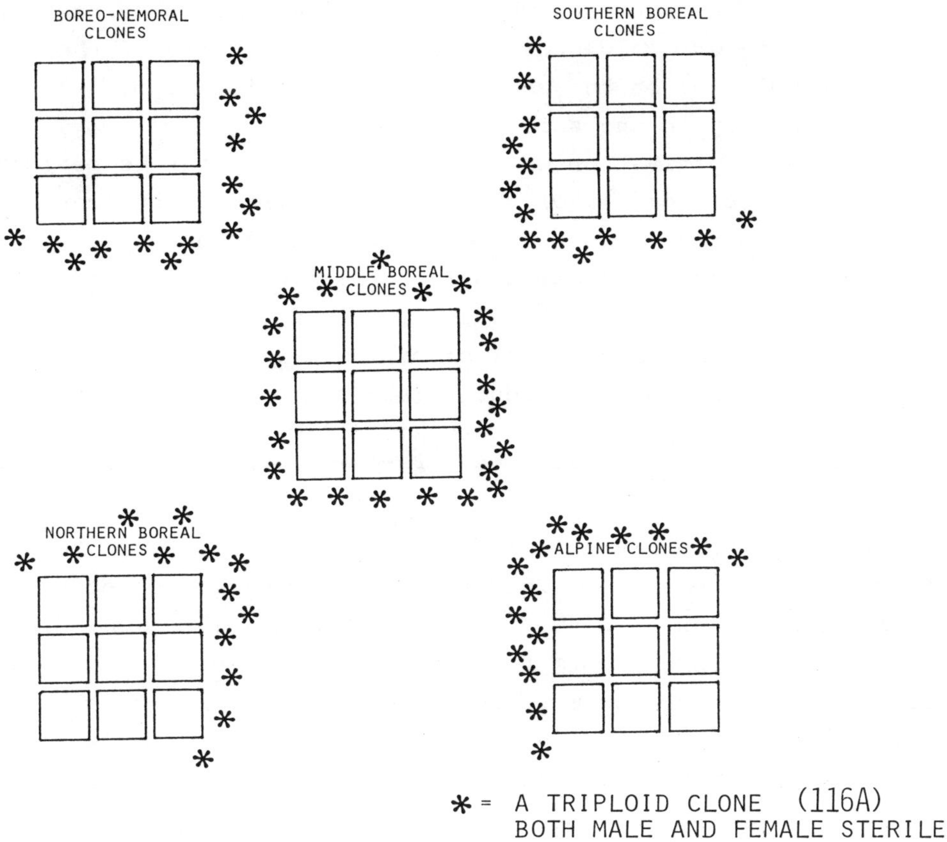

Fig. 5. The structure proposed for the clone collection of the arctic bramble: a subdivision according to the area from which the clones originate

As well as conserving co-adaptation, the bulks described above would also serve to maintain alleles. The maintenance of the bulks in their 'original' milieu should ensure that natural selection will rarely alter gene frequencies so drastically as to cause the disappearance of an allele.

Techniques of collection and storage

Continuous vegetative propagation of the arctic bramble under field conditions is perfectly feasible. Propagation can be accelerated with the aid of mist techniques in the greenhouse, and this is routine practice in the production of planting material of this species. There appears to be little information on long-term cold storage of vegetative parts of the arctic bramble, or on the ability of dormant root stolons to retain viability in the soil.

Meristem cultivation of the arctic bramble can be performed successfully (Bremer, 1980), but for the method to be feasible for genebank purposes further refinements will be needed.

Certain plant tissues, usually callus or suspension cultures, have recently been successfully preserved in liquid nitrogen. So far the survival of meristem tips has been reported at least in carnation, tomato and potato species, in strawberry and in pea (Henshaw et al., 1980). In the future this method might prove to be a possibility in relation to the arctic bramble.

There is probably no information on the long-term storage of arctic bramble seeds. The germinability of dried seeds is generally poor, and treatment with H_2SO_4, as well as chilling, is required to break dormancy. Moist-stored seeds germinate better but would not seem to be well suited to long-term storage (Larsson, 1957, 1969). With suitable improvements in technique, germinability could probably be brought sufficiently high to maintain a given genetic composition. On the other hand, such improvements may defeat the tendencies of natural selection, since it would appear to be ecologically favourable for a proportion of the seeds to remain dormant and only to germinate in subsequent years. Such variation in the depth of dormancy may well be genetically determined. In the opinion of some workers (Larsson, Kjellqvist, personal communication) seeds of the arctic bramble may persist in the soil in a viable state for several decades, perhaps even for one hundred years.

The storage of arctic bramble pollen has not been thoroughly investigated. Pollen has been stored with desiccant in the refrigerator over the winter but germinability has fallen drastically (Larsson, personal communication).

Present natural populations of arctic bramble often fail to produce any berries at all (Figure 2) so that seed harvest is impossible. It is always possible to collect root stolons, although this requires a good deal of labour.

The advantages and disadvantages of the various techniques of collection and storage may be reviewed in the light of the following points.

<u>How much genetic variation is amassed and maintained per unit of resources expended?</u>

To conserve genetic variation in the form of seed is by far the easiest and cheapest method. Digging up root material, transport and subsequently dealing with the packages requires much labour: a piece of rootstock at least 30 cm x 30 cm must be dug up, packed and sent off immediately. At the destination, the brittle rootstock of the arctic bramble must be recovered intact from the dense mass of roots which can take as long as half an hour per sample. Each mass of roots yields at most a single genotype, while by picking a few berries from various parts of the population, one can recover practically all the alleles of the population. On the other hand, the root samples represent the population structure after natural selection whereas seed samples describe the structure of the succeeding population before natural selection. The seed samples thus include less well adapted genotypes.

Conservation as a collection of clones in the field consumes considerable resources since the clones must be isolated underground from each other, weeds must be kept under control, and the clones must be replanted at intervals. In bulk plots no isolation is needed, but large-scale techniques for raising arctic brambles from seed have not been developed. Such techniques may well succeed with slender resources. The bulk seed is produced with little human effort - self-incompatibility and the agency of bees render hand pollination unnecessary. The bulk not only conserves genotypes, but creates new ones; potentially valuable gene combinations arise continually and are recoverable from the bulk population whenever needed. A collection of clones only preserves the *status quo* - nothing new can be expected from it. Bulks, too, must be renewed every few years.

Conservation through controlled crosses in the arctic bramble, i.e. through intercrossing of clones, appears to be too labour consuming to be practised except on a small scale.

The flowers should be carefully isolated and - in spite of self-sterility - also emasculated, due to the pollination technique. The gain per person per summer is only some hundreds of crosses. Alternatively, one should have plenty of isolation cages, equipped with pollinating insects (bumble bees), and the plants should be maintained in pots, which is not easy, due, for example, to the growth habit.

Meristem cultivation is not unduly expensive once the basic equipment has been purchased (Bremer, personal communication). Provided the life of a plantation could be extended sufficiently, for example by cool or deep-freeze storage, the maintenance of the arctic bramble collection might prove to be economical in spite of the fact that this species is technically difficult to handle on account of the small growing-point.

Long-term storage of seed would be considerably cheaper than continuous cultivation, whether it was a question of bulks or more narrowly based populations. When the material was to be utilised, one would have to accept a delay of a few years before the seed could be grown to material which could be evaluated.

How great is the risk of genetic losses?

After long-term storage, one must resort to sulphuric acid treatment, for example, to induce the seeds to germinate (Rose, 1919). However there is always the risk of destroying the seeds when the technique misfires. Nonetheless, seed can be harvested over several years and in fairly large quantities. In the field, clones cannot readily be cultivated in several replications owing to lack of resources. Thus there is the risk that the less vigorous clones will die out (20 - 30% in the first year after collection). There is probably little danger of alleles disappearing from the bulk provided that a bulk is cultivated in a milieu corresponding to the conditions under which the original clones evolved. It is difficult to assess what difference it makes that the bulks will be maintained under cultivated conditions and not under the original, natural conditions - a few alleles may be lost. In meristem cultivation, there is always the risk of genetic alterations, but the risk can be minimised by the use of replications.

How long is the regeneration interval?

Methods of seed storage in arctic bramble have not been developed as yet, but with a sufficient input of research work we may attain a number of decades. The cold storage of vegetative parts has not been investigated. By maintaining meristem cultures at a suitably low temperature (near zero), it might be feasible to extend the regeneration period, possibly even to two years, although this has not yet been attempted with the arctic bramble (Bremer, personal communication). If it proves to be possible to use the deep freeze with the arctic bramble, the regeneration interval of a meristem culture may be extended into decades. Under continuous cultivation, whether of seed-raised or vegetatively propagated material, the regeneration interval is a few years. Stands cannot be maintained for a longer period since their condition starts to deteriorate as the plants use up all their allotted space.

How do conditions of conservation and regeneration influence gene frequencies, co-adapted gene complexes and individual genotypes?

The genetic content of clone collections is not influenced to any extent by the conditions of conservation or regeneration provided that conditions are fairly favourable, so that no clones are lost. Only mutations and the elimination of chromosomes in certain hybrids can cause alterations in the genotypes. In tissue cultures, alterations in the genome are always a hazard. However, in meristem cultures the risk is lower. In raising material from seed, the poor germinability of seed may give rise to genetic drift.

Selection acts rapidly upon gene frequencies in the population if cultivation conditions differ greatly from the milieu from which the original clones were taken. The alteration can be very large, even within a single generation (Sylven, 1937; Hamrick and Allard, 1972). In such cases some alleles, particularly dominant ones, may be lost entirely. However, in practice alleles may be retained in large populations by virtue of intergenotypic interactions, especially in cases where the reproductive value of the heterozygote increases as the heterozygote becomes more rare (Allard and Adams, 1969).

Co-adapted gene complexes break down or alter in character before long, if selection does not favour their maintenance. Such breakdown will occur when populations are grown under different conditions from those under which the co-adaptation evolved. Therefore, bulk populations should be grown under conditions corresponding to those of the original components, and the population size should be kept sufficiently large (thousands of individuals). On the other hand, claims have been made that the evolution of new co-adapted gene complexes tailored to an altered environment need not take many generations. This would ensure the usefulness and wide adaptability of the bulk material, even though it was not possible to retain exactly the original genetic constitution.

The structure of the population's gene pool will also alter greatly from year to year, even within the same generation. Such alterations are especially large if the individuals are not isolated from each other: the vegetatively aggressive strains will spread at the cost of others, until finally the population will consist for the most part of only a few predominant clones. For this reason it is worth keeping the regeneration interval of bulks moderately low, unless one specifically wants to achieve automatic selection for competitive ability in one of the bulk generations (museum material v. breeder's raw material!).

If one wishes to conserve the genetic structure of the population with as little change as possible, the best method is long-term storage of seed with the regeneration interval as long as possible, since during the regeneration phase natural selection will act in any event.

When the object of conservation is a particularly valuable natural population, it can be conserved in a genetically unchanged state (on account of the small number of clones involved) by maintaining the constituent clones individually. Naturally, the clonal structure of the population sample must be determined in advance. Such a procedure might be necessary in cases where one wished to preserve genetic co-adaptation associated with the special conditions prevailing in marginal areas.

What population size can be considered adequate, and how many populations should be conserved?

Let us assume that it is feasible to conserve alleles, co-adapted gene complexes corresponding to the five zones described earlier, and initially as many different arctic bramble genotypes as possible. The genetic variation is narrow within present-day populations of the arctic bramble (Tammisola, unpublished). Most of the genetic variation is to be found between populations. The relation of the variation to natural geographic areas remains to be established. However, on the basis of the foregoing, it appears that one should try to sample individuals from as many populations as possible. The sample size per population can be fairly small, since due to the population structure (one or two predominant clones), the risk of duplicate sampling of the same clone is great, and the genetic yield increases only slowly with increasing sample size. A sample size of 3 root samples, or a berry (1 berry = 30 seeds) gathered at random from various parts of the population, is probably adequate. In order that the aims outlined above be realised, one hardly needs to contend with internal genetic variation in natural populations. With finite resources, a more representative selection of genotypes is obtained by stepping up the number of populations than by increasing sample size.

How great is the risk of contamination? What sort of isolation is needed?

When grown in the field, different clones will intermingle unless one takes steps to isolate them from each other under the ground. This is a crucial problem in clonal collections. In bulk populations no internal isolation is required. If grown in the same region, different bulks must be isolated from each other, though normally each bulk will be grown in its own region. In clonal collections in the field, the regeneration of seedlings is an ever-present danger. The use of a suitable mulch can prevent seedlings from growing. In bulk populations seedlings do not matter; nor does possible migration in the form of seed or pollen from locally growing wild populations, provided the bulks are established in their 'own' natural geographical areas.

Problems can arise from seeds persisting in the soil and germinating in subsequent years, when entirely different arctic bramble strains may be grown later on the same site. Stolons also pose similar problems because a considerable volume of roots remains in the soil after a stand has been ploughed up. We know practically nothing about the ability of these roots to remain dormant in the soil, perhaps to sprout again much later. To avoid such risks, collections should always be moved to a field where arctic brambles have not previously been cultivated.

In clonal collections of arctic bramble, the transference of pollen between collections representing different areas should be prevented. Even a fairly generous-seeming distance does not guarantee adequate isolation (Johansen, 1968) since honey bees and bumble bees are to a large extent species-specific, sometimes even variety-specific, in their foraging habits (Macior, 1968; Anasiewicz and Warakomska, 1969; Faulkner, 1976, Teras, 1976, Gary et al., 1976, 1978; Ferguson and Free, 1979). Nevertheless, when bees work within a stand they appear to fly only short distances from flower to flower so that pollen 'wears off' quickly (Crane and Mather, 1943; Heinrich and Raven, 1972). In practice, therefore, arctic bramble will not be transported more than a few metres within a stand (Figure 6).

The best means of isolation would be to plant a flowering but sterile stand of the same species. There is a suitable strain of arctic bramble: the triploid clone 116A ($2n = 3x = 21$) collected from the wild. This clone is fully male- and female-sterile, but produces both flowers and nectar in abundance (Figure 5).

<u>How can pests and diseases be kept under control</u>?

With the aid of meristem culture and the special techniques associated with it, disease-free 'clean' arctic bramble material can be produced (Bremer, 1980). As long as clean plants are conserved as meristem cultures, renewed infection is impossible. However, the culture medium may sometimes become contaminated and therefore it is worthwhile to duplicate tissue cultures.

Disease does not usually present any problems in bulk populations. Certain diseases are less readily transmitted by seed than vegetatively, especially if the seeds are dressed.

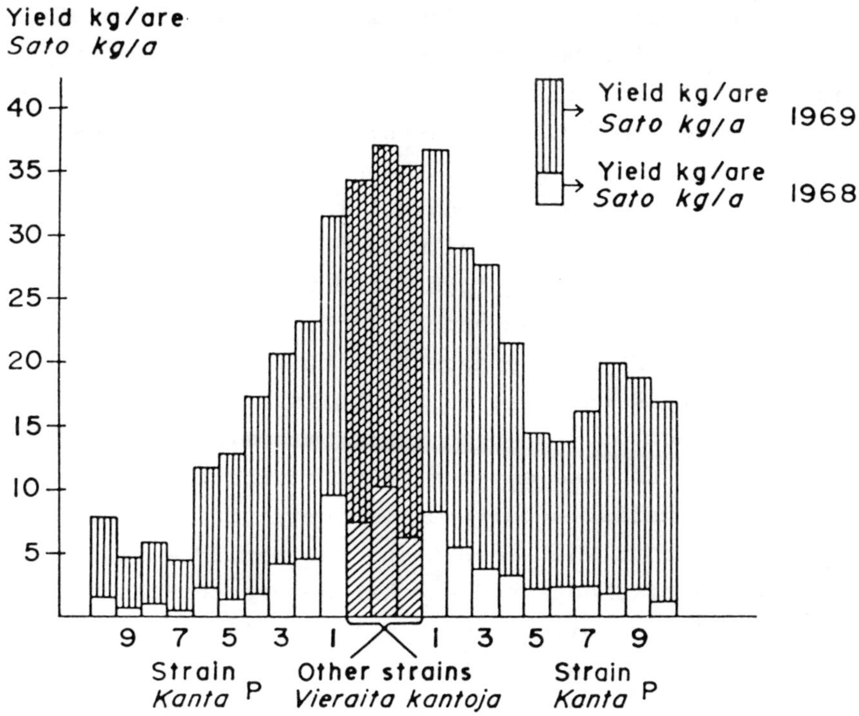

Fig. 6. Yields of an arctic bramble strain at various distances from other strains in 1968-69 (Ryynänen, 1973)

The heterogeneity of the population (also in respect of resistance genes) will hinder the spread of diseases, and natural selection directed at sick individuals will ensure the maintenance of equilibrium between host and pathogen.

When a clonal collection is grown in the field for any length of time, fungus (Ruokola, 1981) and virus diseases start to gain a hold. In the arctic bramble, three rust and one mildew species may exist, in addition to many mildly pathogenic soil fungi. The clones should be cleaned at intervals by the appropriate techniques. The clones should be healthy particularly when evaluations are being made since in other species, the raspberry for example, healthy stands can produce up to four times the yield of virotic ones (Bremer, personal communication).

It should be easier to control pests, diseases and weeds in genebank collections than in commercial cultivation, since on the one hand, the aims are more limited (prevention of genetic losses) whilst on the other hand, the product is not marketed for consumption.

As yet there are not many pests; presumably the most troublesome ones are the thrips, the beetle *Anthonomus rubi* and a bug *Capsus ater*. When chemical control is carried out on bulks due regard must be given to the indispensable bees. Weeds are a problem in arctic bramble plots since this species is a poor competitor and does not even appear to tolerate herbicides (Ryynänen, 1973). It may be necessary to resort to a greenhouse during the seed-raising phase of the bulks.

During collection, the best way of minimising the risk of diseases would be to take shoot samples of the clones for the establishment of meristem cultures in quarantine. At present, little is known about what conditions the method would impose on the condition of shoots and whether such conditions could be fulfilled during the collection of material far from culture facilities. An easier alternative, although involving a greater risk of introducing diseases, is to harvest as seed. Pathogens on the surface of the seeds can be controlled by seed dressing but internally transmitted seed pathogens will become apparent only after germination, in the plants themselves.

What kind of observations can be incorporated into the technique?

The most important observations on populations and individuals can, and should, be made during collection, irrespective of the technique. Since clones cannot generally be distinguished in natural stands, clonal observations are not worth attempting.

Observations can be made every year of plants continually growing in the field. In the case of long-term storage, for example as seeds, at least the most important characters should be observed during the regeneration phase - no evaluation is possible during the storage phase.

In clones conserved as meristem cultures, the most basic characters of the arctic bramble - aroma, yield level, growth habit, disease resistance and other associated factors - can scarcely be evaluated. All these characters can be observed on clonal collections in the field.

On bulk populations, it is quite feasible to make observations on the population level. If the bulk population is dense (established by sowing) it will be practically impossible to distinguish individual clones, or to make individual observations on them. Observations can be made only on individual stems, so the degree of reliability gained is exceedingly low; nonetheless, it may be adequate for the breeder making a rapid pre-screening. Promising looking stems can be dug up and replanted, and selection carried out properly when the plants have grown into small clonal colonies. On the other hand, if the bulk population is sufficiently sparse (established by planting), clonal observations can be made during the initial stages although such observations on small young plants can only be preliminary. This type of bulk is more labour intensive and accommodates a lower number of individuals than the drilled one. In any event, observations of direct value for breeding purposes can be made on the bulks.

ACKNOWLEDGEMENTS

I wish to thank Professor P.M.A. Tigerstedt for the valuable conversations and comments he has contributed to this work. My thanks are also due to Mr. Peter Joy for his valuable contribution in translating this manuscript.

REFERENCES

Allard, R.W., 1975. The mating system and microevolution. Genetics 79, 115-126.

Allard, R.W. and Adams, J., 1969. The role of intergenotypic interactions in plant breeding. Proc. XII Intern. Congr. Genetics, Vol. 3, 349-370.

Anasiewicz, A. and Warakomska, Z., 1969. Occurrence of bumble bees on alfalfa *(Medicago media* pers.*)* in the province of Lublin and pollen analysis of their pollen loads. Ekologia Polska (A) 17, 587-609.

Anon, 1977. Naturgeografisk regionindelning av Norden. NU B 1977, 34.

Bremer, K., 1980. Produktion av friska kärnplantor av bärväxter. Nordisk plantekernkonferance. 1. -7. March 1980. Noresund, Norge, 106-112.

Crane, M.B. and Mather, K., 1943. The natural cross-pollination of crop plants with particular reference to the radish. Ann. Appl. Biol. 30, 301-308.

Faulkner, G.J., 1976. Honeybee behaviour as affected by plant height and flower colour in Brussels sprouts. J. Apic. Res. 15, 15-18.

Ferguson, A.W. and Free, J.B., 1979. Production of a forage-marking pheromone by the honeybee. J. Apic. Res. 18, 128-132.

Focke, W.O., 1910-1914. Species Ruborum. Monographie Generis Rubi Prodromus, 3 vols., Stuttgart.

Gary, N.E., Witherell, P.C. and Marston, J.M., 1976. Inter- and intra-orchard distributions of honeybees during almond pollination. J. Apic. Res. 15, 43-50.

Hamrick, J.L. and Allard, R.W., 1972. Microgeographical variation in allozyme frequencies in *Avena barbata*. Proc. Nat. Acad. Sci. USA, 69, 2100-2104.

Heinrich, B. and Raven, P.H., 1972. Energetics and pollination ecology. Science 176, 597-602.

Henshaw, G.G., Stamp, J.A. and Westcott, R.J., 1980. Tissue cultures and germplasm storage. In: Plant Cell Cultures: Results and Perspectives. Ed. F. Sala, B. Parisi, R. Cella and O. Ciferri, 277-282.

Hiirsalmi, H., 1974. Mesimarjan marjontaan vaikuttavista tekijöistä. (Factors affecting fruit setting in *Rubus arcticus*). Puutarha 74, 626-627.

Hulten, E., 1968. Flora of Alaska and neighboring territories. Stanford University Press, California.

Hulten, E. 1971. The circumpolar plants. II. Dicotyledons. Kungl. Sv. Vet. Acad. Handb. 4. Ser. 13,1: 164-165, 373-375.

Hardh, H. 1976. Puutarhanviljelyä (Horticulture). An interview by U. Iivarinen. Maailma ja Me 11, 166-169.

Johansen, B.R., 1968. Isolation distances in lucerne. In: Royal Veterinary and Agricultural College Yearbook, 1968, 153-158.

Kallio, H., 1975. Identification of volatile aroma compounds in arctic bramble, *Rubus arcticus* L. and their development during the ripening of the berry, with special reference to *Rubus stellatus*. Sm Diss. Turku.

Knight, R.L. and Keep, E., 1958. *Rubus* L. II Floral biology and seed formation. In: Handbuch der Pflanzenzüchtung. Ed. H. Kappert and W. Rudorf, Parey, Berlin 1962, 6, 483-487.

Kotimäki, M. and Hiirsalmi, H., 1979. Cytogenetic studies on *Rubus arcticus*, *R. stellatus* and their hybrids. Hereditas 91, 83-89.

Larsson, E.G.K., 1955. Odlingsförsök med akerbär, *Rubus arcticus*. Sver. Pomol. För. Arsskr., 183-195.

Larsson, E.G.K., 1957. Tetraploid *Rubus arcticus*, framställd genom colchicinbehandling. Bot. Not. 110, 151-159.

Larsson, E.G.K., 1969. Experimental taxonomy as a base for breeding in northern Rubi. Hereditas 63, 283-351.

Macior, L.N., 1968. *Bombus* (Hymenoptera, Apidae) queen foraging in relation to vernal pollination in Wisconsin. Ecology, 49, 20-25.

Poppius, B.R., 1903. Blombiologiska iakttagelser. Acta. Soc. F. Fl. Fenn 25.

Rose, R.C., 1919. After-ripening and germination of seeds of *Tilia*, *Sambucus* and *Rubus*. Bot. Gaz. 67, 281-308.

Ruokola, A.-L., 1981. Fungus diseases of cultivated arctic bramble (*Rubus arcticus* L.) in Finland. J. Scient. Agric. Soc. Finland, 53, 83-89.

Ryynänen, A., 1972. Arctic bramble (*Rubus arcticus* L.) a new cultivated plant. Ann. Agric. Fenn. 11, 3: 170-173.

Ryynänen, A., 1973. *Rubus arcticus* L. and its cultivation. Ann. Agric. Fenn. 12, 1-76.

Saastamoinen, S., 1930. Mesimarja (*Rubus arcticus* L.) Suomessa. Referat: Die nordische Himbeere (*Rubus arcticus* L.) in Finnland. Ann. Zool. Bot. Soc. Fenn. Vanamo, 13, 356-414.

Silen, F., 1906. Blombiologiska iakttagelser i Kittilä, Lappmark. Meddel. Soc. F. Fl. Fenn, 31.

Sylven, N., 1937. The influence of climatic conditions on type composition. Imp. Bureau Plant Genetics, Herbage Bull. 21, 8 pp.

Tammisola, J. and Ryynänen, A., 1970. Incompatibility in *Rubus arcticus* L. Hereditas 66, 269-278.

Teräs, I., 1976. Flower visits of bumblebees, *Bombus* Latr. (Hymenoptera Apidae), during one summer. Ann. Zool. Fennici, 13, 200-232.

Vaarama, A., 1949. Cytogenetic studies on two *Rubus arcticus* -hybrids. J. Sci. Agr. Soc. Finland, 20, 67-79.
Warming, E., 1886. Om nogler arktiska soerters biologi. Bih. t. Kungl. Svenska Vetenskapsakad. handb. 12. Afd. III.

DISCUSSION

G. de Bakker *(FAO)* It seems that this is a typical Nordic/Arctic species; is there any chance, now that you have succeeded in domesticating it, that it can be grown in other areas farther south? How far south do you think it can go?

J. Tammisola *(Finland)* There are some possibilities with the sub-species *stellatus*, from North America, crossed with *arcticus*. However, even for this cross species I think southern Sweden would be about the limit. *Arcticus* itself cannot be grown farther south than southern Finland.

E. Porceddu *Italy)* I think this is a good example of the problems posed by a species which is not threatened by extinction or genetic erosion. It shows that the problems are similar in different fields. When a species is threatened by extinction or genetic erosion the problems are very similar.

Problems of pollination in seed regeneration of cross-pollinated vegetables

L.QUAGLIOTTI
University of Turin, Italy

There are two aspects of this subject, which are in some ways the opposite of each other and which need to be examined with great care in seed growing:

a) the contamination of the product by extraneous pollen;

b) the increased yield that can be obtained by adequate intervention of pollinators or, on the other hand, the reduction in yield which will occur if there are not enough insects.

CONTAMINATION

It is very difficult to assess the extent of contamination in air pollinated plants, as it varies according to environmental conditions and how far the pollen can be transported. With regard to this second factor the range of variability of vegetables is not wide, since only swiss chard or spinach beet (*Beta vulgaris* L. var *cycla* L. Ulrich), red garden beet (*Beta vulgaris* L. var. *esculenta* L.) and spinach (*Spinacia oleracea* L.) are wind pollinated crops.

With regard to the direction, speed and frequency of wind, each area obviously has its own characteristics. Nonetheless the relationship between contamination and the distance between the crops is clear and constant: a considerable reduction in contamination can be obtained by increasing this distance from 3 to 6 m. Hence, in practice, the requirements of spatial isolation for air pollinated plants become very similar to those for insect pollinated crops.

This is so, in practice, even though, in theory, there are the following essential differences between the two pollination mechanisms (Bateman, 1947a, b):

- pollinating insects mostly work independently of one

another but in a predictable way in terms of statistics; movements of large masses of air, due to modifications in the direction and speed of the wind, are not so easy to foresee;

- pollinating insects carry pollen systematically from one flower to another of the same species, whereas wind-carried pollen is dispersed in all directions and settles on flowers of all kinds, according to chance;
- insects' carrying power is limited; hence the quantity of pollen taken from one variety of flower can only increase at the expense of that of another variety or species, whereas there is no limit to the extent of air pollination and the concentration of one type of pollen does not directly affect that of other types. From the practical point of view the problem changes when it is a question of
  i) commercial production of seed or
  ii) obtaining small quantities of seed necessary in the course of breeding, or regenerating stocks etc.

Commercial seed production

Contamination by pollen from species or cultivars different from those to be multiplied must be considered not only in terms of frequency but also with regard to the gravity of the damage it causes. That is to say, there are different degrees of gravity in the undesirable effects of cross-pollination. This situation has been taken into account in the rules (29 September 1970) established by the EC with regard to minimum isolation distances.

The text speaks of "sources of foreign pollen which can cause serious deterioration" and of "other sources of foreign pollen affecting cultivars . . . .". That is, cross-pollination may be particularly serious if the effect is obvious even in the first generation and concerns characters of merceological interest: the colour of chicory leaves, for example, or the form of a lettuce head, etc.

In this connection it should be noted that the minimum

distances prescribed by law (Table 1) do not solve the problem, since they merely represent the minimum threshold that can realistically be accepted, particularly in areas which specialise in such crops (such as Emilia-Romagna and the Marche, in Italy). In fact, these distances do not guarantee satisfactory isolation; distances constituting a real precaution might be somewhere between 1 000 and 3 000 m for the anemogamous species and between 500 and 1 000 m for entomogamous ones.

TABLE 1

MINIMUM DISTANCES (m) BETWEEN CROPS FOR THE PRODUCTION OF BASIC AND CERTIFIED VEGETABLE SEED (EC, 29.9.1970; ITALY 20.4.76, n. 195)

Species	Basic seed	Certified seed
1) Beta and Brassica		
When the foreign pollen can cause serious deterioration	1 000	600
From other sources of foreign pollen	500	300
2) All other cross-pollinating species		
When the foreign pollen can cause serious deterioration	500	300
From other sources of foreign pollen	300	100

In cases where these distances cannot be respected and in the areas that specialise in vegetable seed production, the most effective system is for the local authorities to mark out zones which are exclusively for the reproduction of certain cultivars. In this connection it is important to mention the recent intervention of the Emilia-Romagna Region* concerning beets and onions and of the Marche Region** concerning beets and cabbages.

*Legge regionale n.30 del 6.7.1977 "Norme per la produzione di sementi di piante allogame" (Bollettino Ufficiale Regione Emilia-Romagna n.99 del 7.7.1977). Regolamento di esecuzione della legge regionale 6.7.1977 n.30, concernente la produzione di sementi di piante allogame, 10.3.1979 n.6 (Bollettino Ufficiale Regione Emilia-Romagna n.27 del 14.3.1979).
Legge regionale n.47 del 2.6.1980 "Integrazione e modifiche della legge regionale 6.7.1977 n.30" (Bollettino Ufficiale Regione Emilia-Romagna n.91 del 4.6.1980).

**Legge regionale n.23 del 20.6.1979 "Norme per disciplinare la produzione di sementi di piante allogame". (Bollettino Ufficiale Regione Marche, n.34 del 22.6.1979).

With regard to Emilia-Romagna, the law makes provision for the setting up, for a specified time, of closed areas for cross-pollinated, and non cross-pollinated plants which need to be isolated for genetic and phytosanitary reasons. A special commission, appointed by the Region, establishes the growing priority in favour of one variety or group of varieties. The Region contributes up to 35% of expenses for work carried out to protect healthy and desirable growing conditions within the closed areas. Varieties of groups other than those which enjoy the right of priority may be grown, provided they are at distances greater than those shown in Table 2. If these distances are not respected, the crops that do not have the right of priority have to be eliminated.

Lastly, in commercial growing crops, contamination by pollen from wild plants growing near the seed crops (in meadows, on wastelands, etc.) may also be serious, as may the contamination of the plants growing wild in the seed crops themselves. This is particularly serious in our country, where there are a number of wild species compatible with the cultivated vegetables: carrot chicory, lettuce, fennel, etc. Turnips and radishes may be contaminated by pollen from *Brassica rapa campestris*, *Raphanus raphanistrum*, etc.

Breeding

In this work and conservation of germplasm, cryptic contamination is the most harmful. Cryptic contamination is that which does not become apparent in the first generation (e.g. where a recessive allele contaminates a homozygous dominant genotype) or which does not concern very obvious characters.

On the other hand it is clear that the tolerance acceptable in the case of crops grown for commercial seed production is out of the question for the stocks, in which absolute genetic purity is necessary.

If spatial isolation cannot be achieved satisfactorily, other fairly cheap precautions can be used, such as isolation barriers created by strips of vegetation (hemp, sunflowers, etc.) which keep back some of the wind and draw insects away from the crop to be isolated; strips round the plot, of the crop

TABLE 2

LAW OF EMILIA-ROMAGNA REGION ON THE SEED PRODUCTION OF CROSS-POLLINATING SPECIES. MINIMUM DISTANCES FOR BEETS AND ONIONS.

BEETS	
Sub-species	Groups
Sugar beet	
Fodder beet	
Garden beet	Red flat root Pinkish flat root Red round root Pinkish round root
Swiss chard	White midrib and green blade White midrib and yellowish blade Green midrib and green blade
Minimum distances (m)	
Between cultivars of different groups in the same sub-species	600
Between cultivars of different sub-species	1 000
For pre-basic and basic seed production:	
Between multigerm and cultivars of the same group	600
In all other cases	1 000

ONION	
Groups	
Yellow globular bulb	
Yellow flat bulb	
Red globular bulb	
Red flat bulb	
White globular bulb	
White flat bulb	
cv. 'dorata di Parma'	
cv. 'ramata di Milano'	
cv. 'bianca di Lisbona'	
Minimum distances (m)	
Between cultivars of the same group	100
Between cultivars of different groups	300
For pre-basic and basic seed production:	
Between cultivars of the same group	300
Between cultivars of different groups	500

itself, which are removed at harvest time and whose function is the same as that of the isolation barriers; increasing the area of the plots reserved to a single cultivar or sample.

In any case the safest method is to use isolator cages. There is such a wide range of materials available today which can ensure this effect. There are however two disadvantages of this system: the cost of the structures and their installation, and the modifications of the phenotype that may be induced by the micro-environment with less light and greater humidity (hence particularly favourable to attacks from aphids), etc., especially if the isolator cage has to remain over the plants for rather a long time (e.g. 4 - 6 weeks in the case of the carrot or pepper).

INCREASED YIELD

The increased yield that can be obtained by sufficient pollination is by no means a negligible aspect of seed production. Whereas years ago pollination was a normal, natural event, today it presents a problem which may condition the success of many crops of fruit vegetables (strawberries, cantaloupe, beans, etc.) and not only that of seed crops.

Commercial seed production

The lack of pollinators is often very serious for many reasons: the greater areas given over to specialised crops; the often reckless use of pesticides which destroy a great many insects, and above all bees, which are excellent pollinators; the destruction of wild vegetation, a natural source of pollen and nectar during the periods when the seed crops are not yet in flower; the very widespread use of weed-killers which destroy uncultivated areas where spontaneous pollinators make their nests; the breeding of new cultivars that are difficult to pollinate or even of hybrids which necessarily require intense cross-pollination.

When new commercial varieties are bred, the requirements of cross-pollination are not generally taken into account. Sometimes the nectaries are small and function poorly, the nectar produced is not very attractive, the kind of dehiscence of the anthers does not favour the activity of the insects in that area, etc.

The aim should be that the pollen is carried more and more by insects, since theirs is the least expensive labour. To this end, some important precautions should not be overlooked. Any competitive wild plants that may be growing near the seed crop must be removed in time; never plant the crops themselves near others more attractive, above all to honey bees (for example vetch or lucerne as compared with onion). In fact the presence of extraneous pollen creates competition between cultivated and wild plants (*Taraxacum, Raphanus* clovers, *Veronica,* etc.) in attracting pollinators, very often already insufficient in number. Cucumbers, particularly the cultivars used for pickling, are subject to very great competition both from other cultivated species and from wild plants (Table 3).

TABLE 3

WILD AND CULTIVATED PLANTS MORE ATTRACTIVE TO HONEY BEES THAN CUCUMBERS. (COLLISON AND MARTIN, 1970)

Highly competitive	Weakly competitive
Melilotus alba	*Polygonum* sp.
Melilotus officinalis	*Daucus carota*
Medicago sativa	*Vicia villosa*
Leonurus cardiaca	*Cichorium intybus*
Asclepias syriaca	*Brassica* sp.
Tilia americana	*Linaria vulgaris*
Aretium minus	*Chrysanthemum leucanthemum*
Sonchus arvensis	*Asclepias tuberosa*
Centaurea maculosa	*Erigeron* sp.
Trifolium hybridum	*Sambucus canadensis*
Helianthus annuus	*Monarda fistulosa*
Solidago sp.	*Potentilla recta*
Trifolium repens	
Taraxacum officinale	
Cirsium arvense	
Cirsium lanceolatum	
Nepeta cataria	
Cucumis melo	
Citrullus vulgaris	
Cucurbita sp.	
Cuburbita pepo	
Lotus corniculatus	
Rhus glabra	
Lythrum salicaria	

Onion, too, is very unattractive to bees; above all, it appears, because of the high potassium content in the nectar. So much so that breeders endeavour to reduce the quantity of this element in the nectar of the flowers of the parent lines of hybrids; besides which, the honey bees themselves might be pre-conditioned to accept high potassium levels by being fed with prepared sweet sugary solutions rich in this substance (Waller et al., 1972; Shasha's et al., 1973; Carlson, 1974).

Obviously it is also very much worthwhile for seed growers to keep in mind many points about the new bee-keeping techniques, so as to obtain the greatest possible contribution from pollinators. These include factors favouring mobility of colonies, concerning mechanisation of the operations, concentration, distance and arrangement of the hives, training bees to fly towards a certain source of nectar, etc.

Regeneration of small stocks of seed of high genetic value

This, generally, necessitates working under isolators. Thus, in cross-pollinated plants, pollination must be effected either manually (by brushing the pollen on) or by insects specially bred in captivity.

Concerning the effect of the former, the preliminary results obtained on okra (*Abelmoscus exulentus* L. Moench) are indicated in Table 4. The newly introduced crop was cultivated for the second year in 1980, in our experimental field of Schierano in North Italy. The great effect of hand pollination on seed setting, even where open pollination is concerned, is probably due to scarcity of pollinators visiting this species, in an area where it had never been grown before.

TABLE 4

EFFECT OF POLLINATION ON SEED SETTING IN OKRA

	Number of seeds/ plant
Open pollination	314
Under isolator cage	324
Under isolator cage, with brushing	491

Among the insects, the best pollinators are honey bees, when the size of the isolation cages is sufficient to allow them to survive long enough. The results (Table 5) obtained at Shierano in 1980, in tests on *Phaseolus coccineus*, 'Bianco di

Spagna', are interesting. The plants (54/ha lot) were grown under isolators of 3 x 3 x 3 m; a small colony of about 20 000 bees in a hive (45 x 22.5 x 30.8 cm) with two openings was so arranged that if necessary the bees could also be allowed to gather pollen outside the isolator.

TABLE 5

EFFECTS OF HONEY BEES POLLINATION ON SEED YIELD IN *Phaseolus coccineus* L.

	Number pods/plant	Seeds/plant no.	Seeds/plant g	Number seeds/pod
Open pollination	24	53	87	2.2
Under isolators with few bees	15	31	49	2.1
Under isolators with many bees	30	63	100	2.1

Since, for technical reasons, we were not completely successful in preventing the bees from getting into the control isolated cages, we have no data from this experiment on seed setting without insects. It is however clear that a large number of bees can considerably increase the yield per plant, both in number of pods and in seeds, as compared with what can be obtained in the open air. Consequently, in the area in which we were working, the spontaneous visiting insects actually proved to be insufficient to guarantee the highest yield of seeds. It can also be deduced from the results that it is not so much the quantity of pollen carried onto the single flowers, as the number of flowers visited, that can affect the yield per plant. It is in fact the number of pods and not the number of seeds per pod that varies considerably.

For small isolation cages (even 0.5 m^3) blowflies (*Calliphoridae*) are advisable. They are readily available and easy to handle. They can be bred continuously on waste butcher's meat or can be purchased as larvae intended as fish bait. They cost very little (2 500 lire/kg, the equivalent of about 14 000 larvae); but above all the pupae which can easily be obtained from the larvae, can be stored for 2 - 3 weeks if kept at 7 - 10°C, or for several months if kept at 3 - 5°C; or else they can be turned into adults in 5 - 6 days, if kept at 26 - 28°C.

They do very well on carrots, as appears (Table 6) from the results obtained in Schierano during 1976 and 1977 which, besides indicating a noticeable increase in the yield of seed when *Calliphoridae* were present, show that given comparable types of cage (C<u>v</u>B), seed production can be excellent (D<u>v</u>A) even in isolators.

TABLE 6

EFFECTS OF *Calliphoridae* ON SEED PRODUCTION IN CARROT

	Yield/ plant (g)	% of the yield from the umbels of different orders I	II	III
Open pollination (A)	11	26	49	24
Single plants under isolators (B)	3	7	37	38
Single plants under isolators with blowflies (C)	7	14	56	27
Pairs of plants under isolators with blowflies (D)	12	31	42	17

Moreover it is clear that the presence of the cage modifies the growth habit of the plants and above all the seed setting of the umbels of various orders, whereas it is known that, in natural conditions, the commercial seed comes almost exclusively from the umbels of the I and II orders, as it is advisable to harvest when the latter are ripe.

REFERENCES

Bateman, A.J., 1947a. Contamination of seed crops. II Wind pollination. Heredity, 1, 235-246.

Bateman, A.J., 1947b. Contamination in seed crops. III Relation with isolation distance. Heredity, 1, 303-336.

Bhatnagar, S.P. and Sudhir Chandra, 1976. Reproductive biology of *Abelmoschus exulentus*. VIII, Seed disfunction. Botanische Jahrbücher, Bd 97, 3, 425-430

Carlson, E.C., 1974. Onion varieties, honeybees visitations and seed yield. California Agriculture 9, 16-18.

Collison, C.H. and Martin, E.C., 1970. Competitive plants that may effect the pollination of pickling cucumbers by bees. Am. Bee J., 110, 7, 262.

Faegri, K. and van der Pijl, L., 1971. The principles of pollination ecology. Permagon Press, Oxford.

Free, J.B., 1970. Insect pollination of crops. Academic Press Inc.

McGregor, S.E., 1976. Insect pollination of cultivated crop plants. United States Department of Agriculture, Agric. Handbook No. 496, Washington.

Parthasarathy, V.A. and Sambandam, C.N., 1976a. Studies on self-pollination Techniques in Bhendi (*Abelmoscus exulentus* L. Moench). Annamalai University Research Annual, 6, 76-82.

Parthasarathy, V.A. and Sambandam, C.N., 1976b. Studies on seed yield of local types of Bhendi (*Abelmoscus exulentus* L. Moench). Annamalai University Research Annual, 6, 88-95.

Quagliotti, L., 1977. L'impollinazione nelle colture da seme. Rivista di Agronomia, 11, 1-2, 64-71.

Richards, A.J., 1978. The pollination of flowers by insects. Linnean Society Symposium Series No. 6, Academic Press, London.

Shasha's, N.S., Nye, W.P. and Campbell, W.F., 1973. Path-coefficient analysis of correlation between honeybee activity and seed yield in *Allium cepa* L. J. Am. Soc. Hort. Sci. 98, 4, 341-347.

Waller, G.D., Carpenter, E.W., and Ziehl, D.A., 1972. Potassium in onion nectar and its probable effect on attractiveness of onion flowers to honey bees. J. Am. Soc. Hort. Sci. 4, 535-539.

Woyke, H.W., 1981. Some aspects of the role of the honeybee in onion seed production in Poland. Acta Horticulturae, 111, 91-98.

DISCUSSION

G. Jenkins *(UK)* I think certainly the effect of seed weight from the primary umbel you mentioned in carrots is supported by the work at the National Vegetable Research Station at Wellesbourne. They have similar results.

Y. Demarly *(France)* Since many vegetable cultivars have an unstable equilibrium population, how do you maintain the plants? With carrots, for example, generally the carrot cultivar comes from a few mother plants so if you want to maintain cultivars, you cannot keep going from one generation to the other, you have to come back to the original mother plants. My question is, how do you keep the mother plants?

Luciana Quagliotti *(Italy)* We keep rather a large number of plants - 20, 30 or even 100 plants - under isolation, with blow flies.

Seed regeneration in *Fragaria ananassa* varieties and *Fragaria* x *Potentilla* crosses

KATARZYNA NIEMIROWICZ–SZCZYTT
University of Agriculture, Warsaw, Poland

In recent years, research has been carried out at the Institute of Genetics and Plant Breeding, SGGW-AR, on the determination of sex in *Fragaria* species and *Fragaria* x *Potentilla* hybrids. The work has been under the direction of Professor B. Kubicki.

For this purpose, seeds from varieties, clones and crosses, self- and open-pollinations, and seeds from interspecific and intergenic crosses, were collected.

Female clones comprised the maternal plants in all the experiments. Emasculation of the flower buds of the plants was not necessary as the probability of self-pollination was negligible. The inability to produce functional pollen grains was examined frequently. The following clones of *Fragaria ananassa* were used: Freia, Sonja, Frau Mieze Schindler, Reine des Precose, Dir. Paul Waldbaum, Pozdnaja Słodkaja, Pozdnaja z Kubanii, and Papa Lange. The following hermaphrodite varieties were used as pollinators: Ostara, Redgauntlet, Geneva, Regina, together with other species of *Fragaria* such as *F. virginiana, F. moschata, F. vesca,* as well as various species of *Potentilla.*

Crossing and subsequent collection of seeds was carried out in the years 1977 - 1980 and the germination of seeds from all the parental combinations was examined in 1980. The germination level of the seeds from *Fragaria* x *Potentilla* crosses was estimated for each year.

Crosses were made in the field, isolation of flower buds being effected by tonophan isolators on cotton padding. The fruits were soaked in water until the seeds could be separated. Two weeks after collection the seeds were sown in peat and seedlings were counted at one and two months after sowing.

Abbreviated results of the germination of seeds in various combinations are shown in Tables 1 - 4.

TABLE 1

Fragaria ananassa CROSSES. SEED GERMINATION, 1980.

Cross	Number of seeds	Number of seedlings	% seedlings
Variety crosses			
Freia x Ostara	715	70	9.8
x Redgauntlet	1 120	70	6.3
x Geneva	678	60	8.8
Sonja x Ostara	250	14	5.6
x Redgauntlet	1 778	42	2.3
x Geneva	520	14	2.8
Frau M. x Redgauntlet	354	20	5.6
			x=5.9 9.8-2.3
Open-pollination			
Freia	2 500	190	7.6
Sonja	1 440	98	6.8
Frau Mieze S.	2 040	110	5.3
			x=6.5
Self-pollination			
Ostara	3 440	140	4.0
Redgauntlet	2 540	80	3.1
Geneva	590	17	2.9
			x=3.3

In the clonal crosses of *F. ananassa*, the percentage of germinating seeds did not exceed 10%, the average being 5.9%. In the case of open-pollination of the same clones, it did not exceed 8%, with an average of 6.5%. Self-pollinated hermaphrodite plants gave even lower results, the average being no more than 3.3%. In general the percentage of germinating seeds was very low and suggests a high degree of genetic instability and incompatibility within this species.

It is interesting to compare the data presented above with the results of intergeneric *Fragaria* x *Potentilla* crosses. The percentage of germinating seeds in the successive years did

TABLE 2

Fragaria x *Potentilla* CROSSES. SEED GERMINATION IN 1980. (EXAMPLES)

Cross	Number of fruits	Number of seeds	Number of seedlings	%
Freia x *P. geoides*	263	15 690	48	0.30
x *P. grandiflora*	75	4 010	40	0.99
x *P. fruticosa*	56	2 310	3	0.13
Sonja x *P. geoides*	39	4 046	4	0.09
x *P. grandiflora*	211	15 228	56	0.36
x *P. fruticosa*	88	2 931	17	0.58
Frau M. x *P. geoides*	31	1 907	3	0.15
x *P. grandiflora*	58	2 460	2	0.08
x *P. fruticosa*	40	660	16	2.42

TABLE 3

FRAU MIEZE S. x *Potentilla fruticosa* CROSSES. SEED GERMINATION 1978 - 80.

Year	Number of fruits	Number of seeds	Number of seedlings	%
1978	36	320	1	0.31
1979	16	200	10	5.00
1980	40	660	16	2.42

TABLE 4

Fragaria x *Potentilla* CROSSES. SEED GERMINATION 1977 - 80.

Year	Number of fruits	Number of seeds	Number of seedlings	%	Mature plants
1977	126	2 402	50	2.08	6
1978	781	19 657	128	0.65	39
1979	729	14 222	290	2.03	-
1980	1 110	62 913	296	0.47	near 80

not exceed 2.08%. A minimum of 2 500 and a maximum of 63 000 seeds per year were germinated so that the results are based on a sufficiently large sample.

It is well known that the percentage of germinating seeds depends on factors such as the parents chosen, the temperature and general weather conditions during the flowering and harvesting periods, and the conditions of germination and growth of seedlings. Temperature is of particular importance. It can be seen in Table 4 that the percentage of germinating seeds in each year was different although the number of seeds used was very similar. This applies overall as well as in regard to particular combinations.

It is worth noticing that it is possible to obtain up to 5% of germinating seeds in intergenic crosses which is similar to that obtained with intra-specific crosses, although the general average drops close to 1%. However, for intergeneric crosses the percentage is remarkably high. In practice the percentage of germinating seeds remains at a very low level. It is clear that within such a young allooctoploidal species as *F. ananassa*, incompatibility barriers may exist.

The high seed viability from intergenic crosses can be explained by the embryological experiments carried out by the author using four female clones: Freia, Sonja, Frau Mieze Schindler and Reine des Precose. The flower buds of these clones were observed from the time of nucellus differentiation until fertilisation and embryo development.

The occurrence of false polyembryony, that is of an ovary with two ovules or two nucelli, and nucelli with many embryo sacs, was detected. This has also been observed by other research workers. The embryo sacs can fuse, so the ploidy level of egg cell and of the central nucleus may change.

In the early stages of flowering, buds growing apo- and diplosporic embryo sacs were observed. This also influences the ploidy changes of the offspring. Therefore, the mechanism of apomixis (which without doubt is stimulated by foreign pollination) influences the viability of seeds obtained from intergeneric crosses. It also influences the ploidy of the offspring from these crosses. On one hand, these mechanisms are a specific source of variability but, on the other hand, they complicate and restrict multiplication as well as making genetic research difficult.

It sometimes happens that the interspecific offspring are more viable than those obtained from crosses within the same species (e.g. *F. virginiana* x *F. ananassa*).

In conclusion, it can be said that the classification rules that work within 'old' stabilised species, do not work in research carried out on 'young' species which have retained the characters of hybrids to a high degree.

Session 4
Technical aspects: Field crops

Chairman: H.LAMBERTS

Theoretical aspects and practical implications of cross-pollination on seed regeneration of field crop genetic resources

C.DE PACE & G.T.SCARASCIA MUGNOZZA
University of Bari, Italy

L.M.MONTI
University of Naples, Italy

P.PERRINO, E.PORCEDDU & P.L.SPAGNOLETTI ZEULI
Istituto del Germoplasma, Bari, Italy

INTRODUCTION

Agriculture has undergone a revolution during this century, bringing in its wake new problems for agricultural scientists to solve. One of these concerns the erosion of genetic diversity of important crop plants and their wild relatives. Preservation of genetic variability is essential for future efforts in plant breeding.

Genebanks seek to preserve the still existing variability in crop species. Their functions are:

- to collect and store the seeds under favourable conditions to maintain viability;
- to regenerate seed stocks;
- to characterise and document accessions;
- to distribute seed and information to interested scientists.

Regeneration of seed stocks implies a direct consideration of the genetical characteristics of the species, its reproductive mechanism and environmental effects on these features.

The biological characteristics of self-pollinated crops make the task of handling germplasm collections relatively easy: their seed lots can easily be renewed by harvesting the seeds produced by plants grown without specific care.

Conversely, cross-pollinated crops pose serious problems during regeneration. The genetic integrity of a population is easily lost if plants are not grown in isolation and in a suitable environment, similar to that where they evolved. Factors such as selection pressure, uni-directional gene flow and environmental effect on the mating system could lead the population to a new genotypic structure with the fixation of some alleles and the loss of others.

The aim of this presentation is to gain insight into a proper strategy to regenerate seed stocks of cross-pollinated field crops without losing their genetic identity. In other words it will seek biological conditions which allow the conservation of the gene frequency found at a locus or the integrity of gene complexes, almost at the same level even after 5 - 10 generations of seed multiplication performed in a rational and economic way.

BASIC ASPECTS

Consequences of seed regeneration such as changes in allele frequency and the determination of the number of generations required for an allele to be lost under environmental conditions which differ from those existing where the population evolved, are well described by population genetics.

In practical terms it is not difficult to follow genes during regeneration. It is known, in fact, that major genes have a great impact on the adaptation of crop plants (Qualset, 1979). They can be individually identified through certain important characters such as flowering time, plant height and attributes of canopy structure. In addition there is also much evidence suggesting that genetic variation in continuous characters may often be controlled by relatively few genes (e.g. East, 1916; Payne, 1918; Sax, 1923; Wehahahn and Allard, 1965; Law, 1967; Weber, 1959).

Methods for determining whether the variation in a character is continuous or is due to the action of genes at one or two loci have been proposed by Powers (1963) and Stewart (1969). Approaches for increasing the resolving power of quantitative genetic analysis have been discussed by Qualset (1979). Once marker genes have been identified at a locus which directly controls or is tightly linked to a locus whose genes control fitness characteristics, it will be very easy to follow the rate of change of the frequency of that allele and prevent further changes.

Regeneration of seed stocks while maintaining the costs at reasonable levels requires identification of the smallest population size able to prevent strong dispersive effects on gene frequency, in addition to identifying the conditions for avoiding selective pressures of climatic stresses, pathogens and insects and allowing the maintenance of a proper mating system.

An effort is made here to obtain indications for identifying the factors affecting genetic features of populations during their regeneration. In doing this, the simplest case of genetic systems - a diallelic locus - will be considered and the following conditions assumed:

- mutation occurs at a constant rate of 10^{-6} per generation and its effect on population dynamics has no measurable consequences over a brief period of time;
- reproductive cells are haploid and their union results in a diploid individual. The theory, extended to cases in which other situations prevail - tetraploids, hexaploids, aneuploids and so on (Li, 1976) - shows that, as a general consequence, polyploidy slows the population response to evolutionary forces;
- sex-linked genes will be ignored, because no sex-chromosome has been identified in the majority of crop plants;
- pleiotropy will be ignored, full penetrance and expressivity assumed, and account will be taken of non-cumulative genetic effects within a series of alleles (dominance) and between series (gene interaction or epistatic effects), when proper theoretical models are considered;
- frequency and density dependent selection will not be considered, because seed stocks of plant genetic resources are often multiplied in spaced plantings and competition for resources is usually less.

Factors affecting population structure

A number of factors can produce changes and/or loss of useful genes, during the multiplication of germplasm populations.

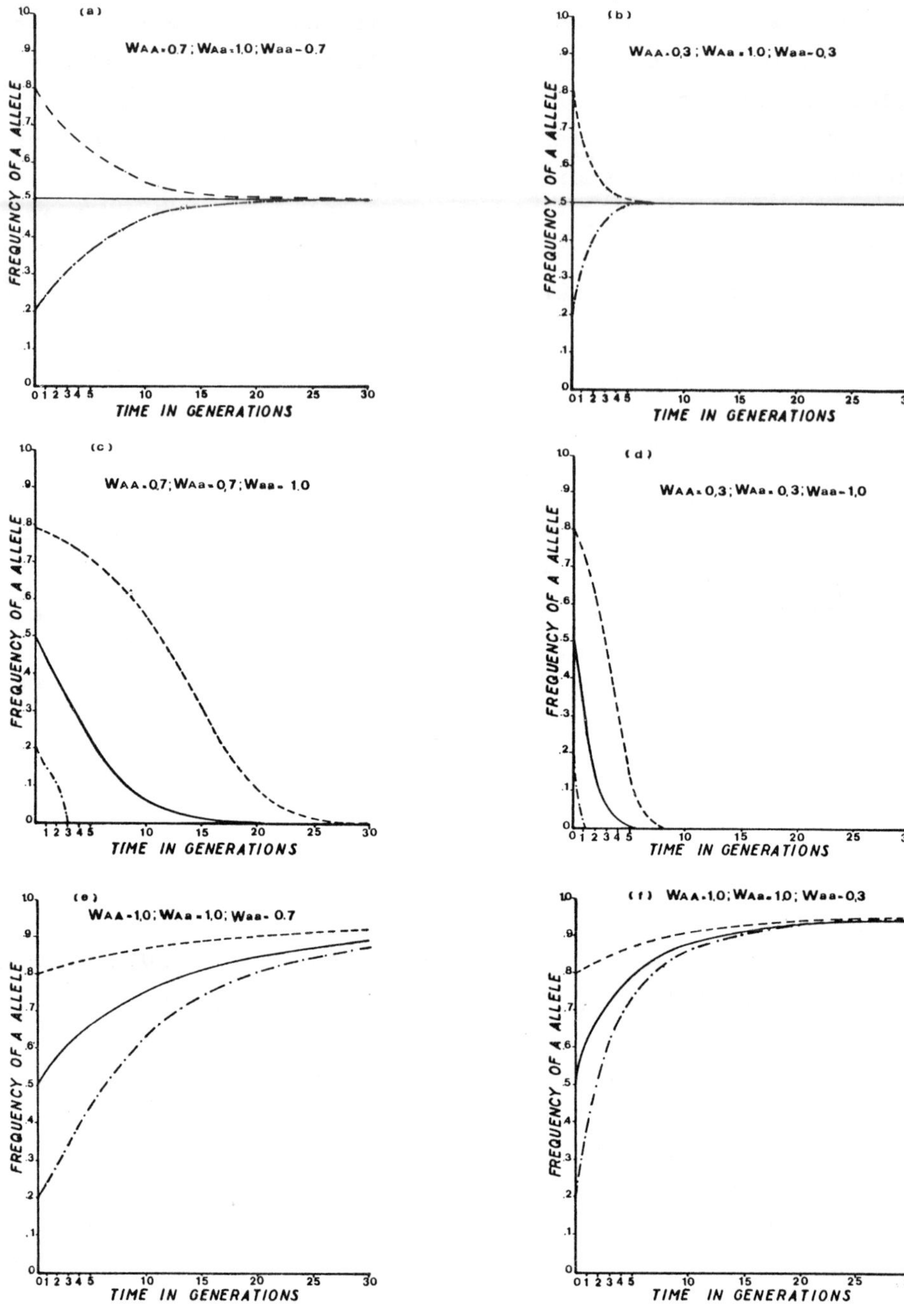

Fig. 1(a) Weak and (b) strong selection against the homozygotes; (c) weak and (d) strong selection against a dominant allele (e) weak and (f) strong selection against a recessive allele. For this figure W_{AA} = fitness value for genotype AA; W_{Aa} = fitness value for genotype Aa; W_{aa} = fitness value for genotype aa the initial conditions at generation 0 were f(A) = p = 0.8 (------); 0.5 (———) and 0.2 (-·-·-·-)

Some of them - fitness, mating system, population size and migration - will be considered in the following points, as a guide to illustrate general principles.

Fitness

Fitness can be affected in various ways, many of which are of interest to genetic resources. Competition among genotypes, differential seed viability, cold and photoperiodic requirements, resistance to environmental factors can all result in an increased or decreased fitness of each genotype.

Figure 1 illustrates the effect of selection on the allele frequency when three different sets of fitness values (simulating the cases of heterozygote advantage, dominant allele favoured, recessive allele favoured) and three different initial allele frequency conditions ($p_A = 0.2$, $p_A = 0.5$, $p_A = 0.8$) are considered.

The allele frequency moves towards an equilibrium ($0 < p > 1$) when heterozygotes are at advantage, or towards fixation ($p = 0$ or 1) when directional selection acts against one of the homozygotes, whatever the initial allele frequency and selection intensities are. The equilibrium allele frequency of $p_A = 0.5$ is reached in almost 25 generations for all initial A allele frequency between 0.2 and 0.8 and for symmetrical selection pressure against the homozygotes of w (Survival rate) = 0.7 (Figure 1a). Stronger selection pressure ($w = 0.3$) against the homozygotes causes the equilibrium allele frequency to be reached earlier (Figure 1b).

When dominance is present, selection against the genotypes carrying the dominant allele causes changes in allele frequency, according to its initial value and selection rate. With mild selection ($w_{AA} = w_{Aa} = 0.7$) it takes three generations to halve the gene frequency ($0.2 \rightarrow 0.1$) when its initial value is low ($p_A = 0.2$); while, in the same period, the frequency is not halved ($0.5 \rightarrow 0.34$) when the initial frequency is higher ($p_A = 0.5$) (Figure 1c).

Heavy selection ($w_{AA} = w_{Aa} = 0.3$), against AA and Aa, makes the situation dramatic because in about seven generations the frequency of A is reduced below 0.05, irrespective of its initial frequency (Figure 1d).

With the same selection pressure ($w = 0.7$ and $w = 0.3$) but acting against the recessive allele _a_, the change in allele frequency is less drastic (Figure 1e). More than 20 generations are necessary to reduce the frequency of the _a_ allele below the critical level of $p = 0.1$, when the initial conditions are $w_{aa} = 0.7$ and $p_a = 0.2$; while it takes nine generations to observe the same decrease of the _a_ allele frequency to 0.1, when the initial conditions are $w_{aa} = 0.3$ and $p_a = 0.2$ (Figure 1f).

Generally speaking, low initial allele frequency ($p < 0.5$) and strong selection ($w = 0.3$) cause the allele frequency to go below the critical value of 0.1 in one generation in the case of the A allele and in nine generations when _a_ is selected against.

Symmetrical selection pressures on the homozygotes allow polymorphism and the equilibrium allele frequency is reached earlier when selection pressures are high.

Mating system

The mating system is one of the strongest factors affecting the genetic structure of the population. The mating system of plant species ranges from obligate selfing to complete outcrossing.

Since the allele frequency does not change in either pure selfing or complete cross-pollination, changes do not occur either in mixed selfing or in random mating. Two opposite forces, namely selfing, which tends to eliminate heterozygotes and random mating, which replenishes the supply of heterozygotes, act at the same time, leading to an equilibrium, in which more homozygotes and fewer heterozygotes occur than with purely random mating.

Genotypic frequencies present in a population when only the mating system is responsible for the genetic variability at one locus are shown in Figure 2. Cases with low ($s = 0.2$), medium ($s = 0.5$) and high degree of selfing ($s = 0.8$) are considered in connection with initial A frequencies of $p = 0.2$, $p = 0.5$ and $p = 0.8$.

When $p_a = 0.2$, (Figure 2a) the aa homozygotes prevail over generations irrespective of the amount of selfing. An increase of _s_ from 0.2 to 0.8 causes a relative increase of the

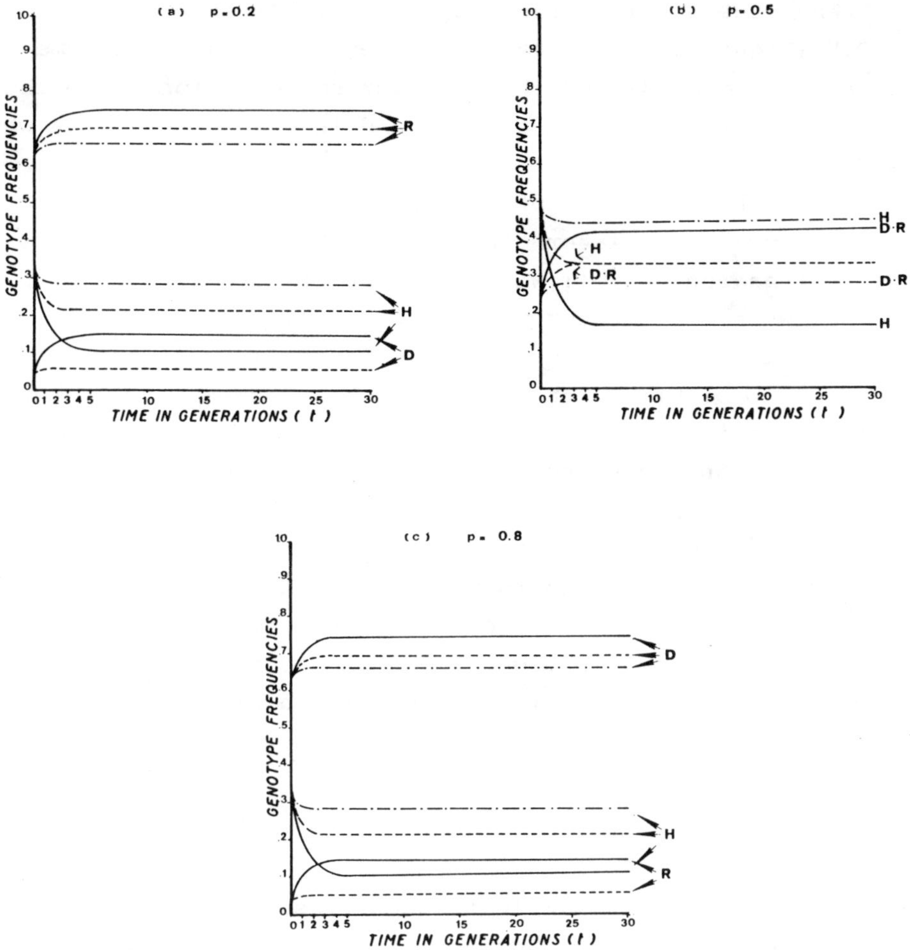

Fig. 2. Genotype frequencies at t + 1 generation when the mating system consists of selfing mixed with random mating. Cases considered are for initial A allele frequency of (a) 0.2, (b) 0.5 and (c) 0.8 and s (= fraction of the population which selfs) 0.2 (-·-·-·-), 0.5 (------) and 0.8 (———). D, R and H indicate frequency trajectories for AA, aa and Aa genotypes respectively.

homozygotes and a proportional decrease of heterozygotes. However, the frequency of AA is very low (< 0.1) when $p_a = 0.2$ and s = 0.2 and 0.5. A selfing rate of 0.8 causes the steepest increase in AA frequency and a decrease in the Aa frequency; after three generations the AA homozygotes will be more frequent than Aa; equilibrium conditions are reached after five generations.

When the initial frequency of A allele is p = 0.5 and s = 0.5 (Figure 2b), the aa and AA genotypes have the same frequency (p = 0.33) and the equilibrium condition is reached in three generations when s = 0.2 and in five generations when s = 0.8.

When the frequency of A allele is p = 0.8 (Figure 2c), it is expected that in later generations the AA homozygotes prevail over aa and Aa following a trajectory symmetrical with that observed when the frequency of A is 0.2.

Selfing rates higher than 0.8 cause frequencies of both AA and aa to be higher than that for Aa. When p = 0.2, random fluctuations in selective values could lead the AA genotype to be lost by chance causing fluctuations in gene frequencies. However, Jain and Marshall (1968) showed that fluctuation in outcrossing rate resulted in a very small change in the critical bounds between situations in which different proportions of homozygotes are present. Stochastic selective values, on the other hand, resulted in significant shifts towards producing a larger region of equilibria with polymorphisms.

The conditions mentioned above produce the following consequences:

a) the probability of extracting homozygous lines is increased;

b) with very low A allele frequency, selfing can cause disappearance of the AA genotypes, unless a large population is grown;

c) high selfing rates could lower the frequency of heterozygotes without changing the gene frequency.

Natural or man-induced modifications of the mating system, in the sense of increasing self-pollination or inbreeding, can speed up the dispersive process. Inbreeding could arise also as a result of a selection pressure favouring the homozygotes. The joint effect of mating system and homozygote advantage can be expressed as the 'fixation index', F, a measure of the lack or excess of heterozygotes compared to those expected in a large, isolated, non-selected, random mating population.

A reduced population size, an increase in the selfing rate or a selective advantage of the homozygotes can increase the number of inbreds over the values expected in a large random mating population. Equilibrium conditions in a population under inbreeding are expected mainly when mixed selfing and random mating and selection are involved. Those conditions have been studied by Hayman (1953) for a diallelic locus (Figure 3). In general it is expected that an increase in the survival rate of the homozygotes and a decrease in the selfing rate can cause a population to have less heterozygote frequency than in an unselected random mating population.

When selection and mating system are factors which determine the genetic structure of the population, the possible dynamics of population starting at time zero at different allele

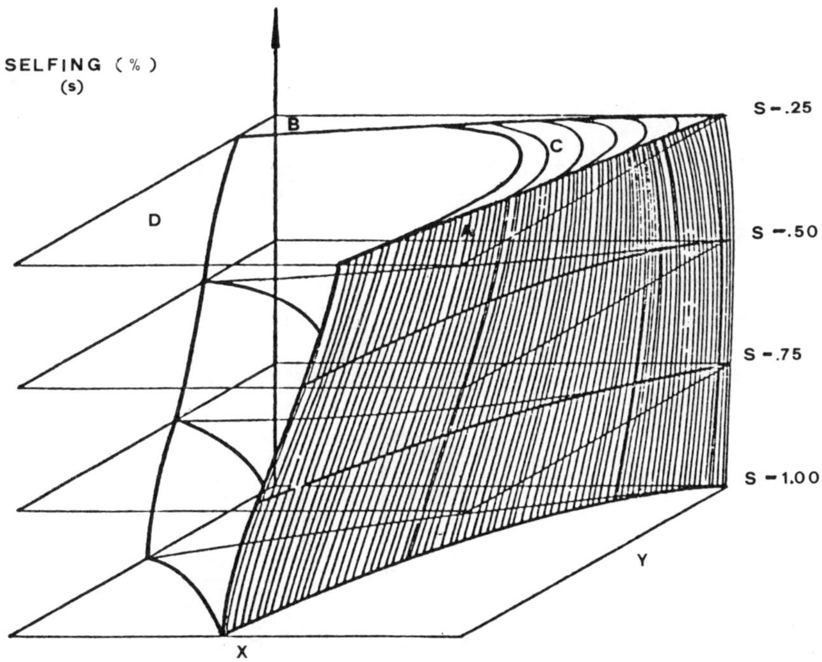

Fig. 3. Phase diagram showing types of equilibrium with various proportions of selfing (s) (adapted from Hayman, 1952).

x = Survival rate of one homozygote (aa) relative to that of the heterozygote (ab).
y = Relative survival rate of the other homozygote (bb).
A = Population becomes homozygous aa.
B = Population becomes homozygous bb.
C = Population becomes less heterozygous than a population mating at random without selection.
D = Population becomes more heterozygous than a population mating at random without selection.

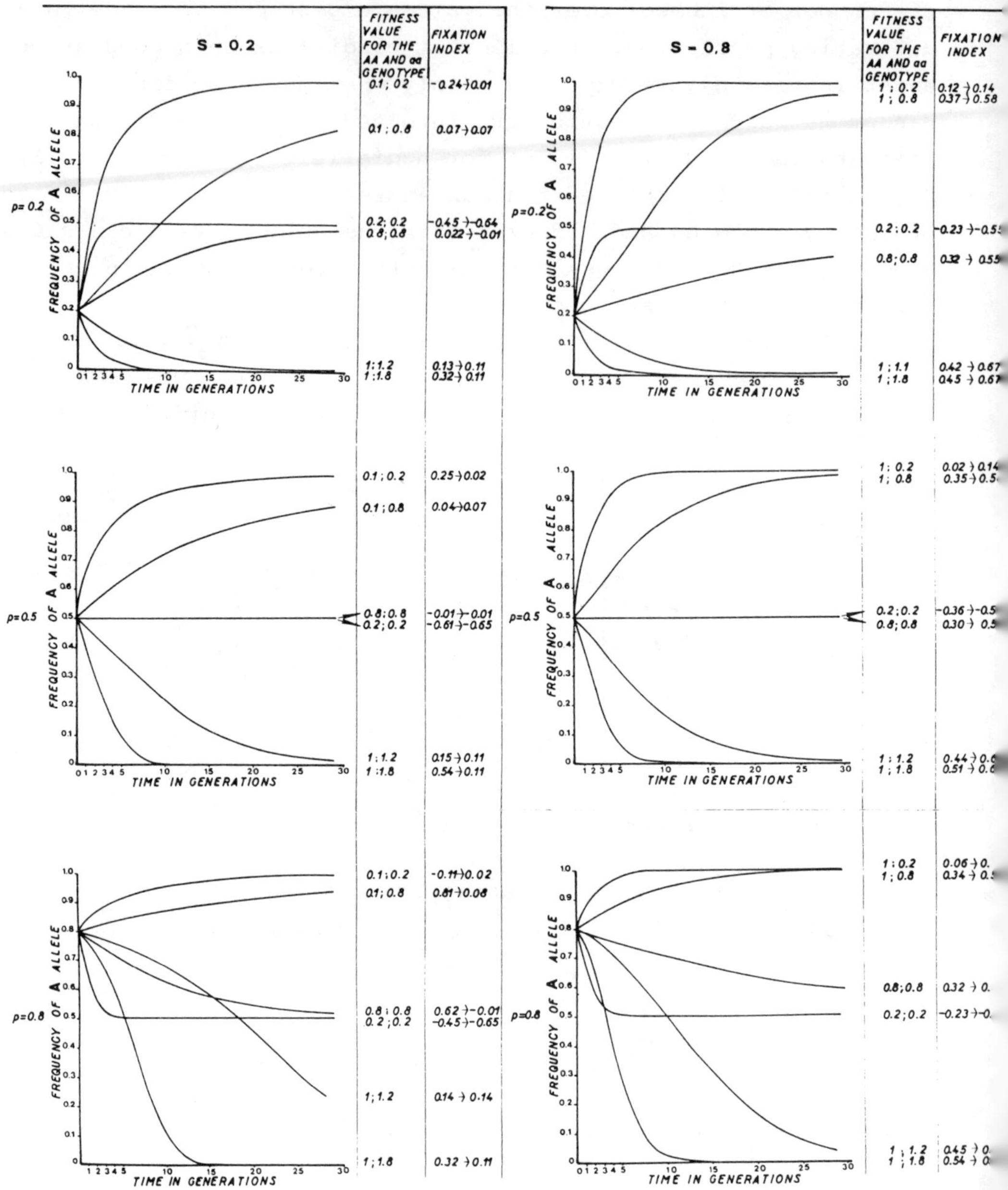

Fig. 4. Diagram showing types of trajectories of gene frequency when various proportions of selfing (S = 0.2, S = 0.8) and random mating (t = 1 - S) and fitness values for genotypes AA and aa are considered three different initial allele frequency (p = 0.2, 0.5, 0.8) are postulated.
X = Survival rate of one homozygote (AA) relative to that of the heterozygote (Aa).
Y = Relative survival rate of the homozygote (aa).

frequency and experiencing both different degree of selfing and survival rate for homozygous are represented in Figure 4. The trajectories of the change in allele frequency are not very different when different degrees of selfing ($s = 0.2$ and $s = 0.8$) are considered.

In a population having low selfing rate (e.g. $s = 0.2$) changes in allele frequency require more time than in a random mating population (see Figure 1). For instance, it takes one generation to eliminate the allele A when its initial frequency is 0.2, and in a situation of strong selection ($w_{AA} = w_{Aa} = 0.3$) and random mating while as many as eight generations are necessary to lose the A allele in the same situation of initial allele frequency and selection regime but with selfing rate of 0.2 (Figure 4); when $s = 0.8$ the loss of the A allele occurs in ten generations (Figure 4d).

The value of the fixation index, F, shows that there is an excess of homozygotes when mixed selfing and random mating are involved. However, it is evident from the F value reported in Figures 4a to 4f that the excess of homozygote genotypes tends to decrease over generations in either cases of selection against the aa or AA when $s = 0.2$, but increases when $s = 0.8$. For symmetrical selection pressure against the homozygotes, a deficiency of homozygotes is expected mainly when fitness is low ($w_{aa} = w_{AA} = 0.2$), while the homozygotes are in excess when mild selection pressure against the homozygotes and high degree of selfing are present.

Population size

Population size often has to be reduced during seed regeneration due to limitations in personnel and availability in resources. In this situation, gene and genotypic frequencies can be subjected to random fluctuations as a consequence of gamete sampling. The final result of random genetic drift related to small population size is the fixation of alleles. This effect often takes several generations but rapid fixation could occur by chance (Figure 5).

This can determine the following consequences:

1) Formation of sub-populations;

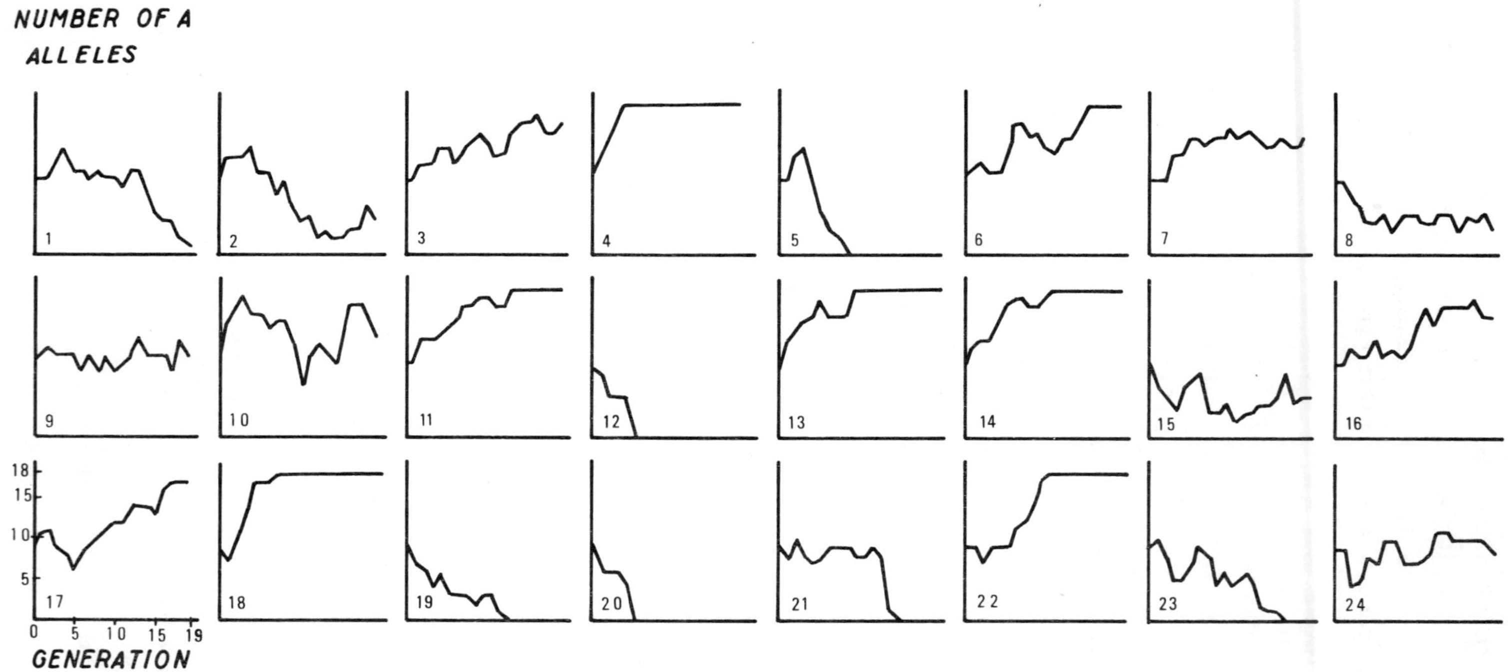

Fig. 5. Change of allele frequency by random genetic drift over 19 generations in 24 hypothetical populations of size N = 9. (From principles of population genetics, by D.L. Hartl, Sinauer Ass., Inc. Publ., Massachusetts).

2) Erosion of genetic variation;
3) Increase of homozygote frequencies.

In small populations, the rate of inbreeding increase is inversely proportional to its initial value. For example, in a population of 20 plants, and with an initial inbreeding coefficient around zero the amount of inbreeding increases to 0.2 in 10 generations (Figure 6a) but it takes 23 generations when the initial inbreeding is 0.5 (Figure 6c).

In a small population (N = 2), inbreeding increases more rapidly than in large populations ($N \geqslant 20$).

The effects due to small population size on the genotypic structure of the population itself can also be estimated as the probability of extracting a heterozygote from the population. The minimum population size, N, in order to have about 50% chance of extracting a heterozygote at one locus at the nth generation is as follows:

Generation:	1	2	3	4	....	5 - 10
N :	3	5	8	10	....	10 - 20

It can be seen that the population size must increase by about two units every generation, until the tenth generation, if a high probability of maintaining heterozygosity at a locus is desired. Other implications are:

- the probability of extracting a heterozygote at one locus decreases as the number of generations increases and population size decreases (Figure 7);
- the minimum number of plants necessary to maintain a probability around 0.5 of extracting a heterozygote at one locus, is 3 in the first generation and 10 to 20 from the 4th to the 10th generation.

Migration

Gene frequencies may also be changed by immigration of genes from other populations. In plants, migration or gene flow can take place as seed flow or pollen flow. The first case can be quite easily prevented during seed regeneration by using accurate sowing techniques and threshing the seeds with self cleaning inspectable machines.

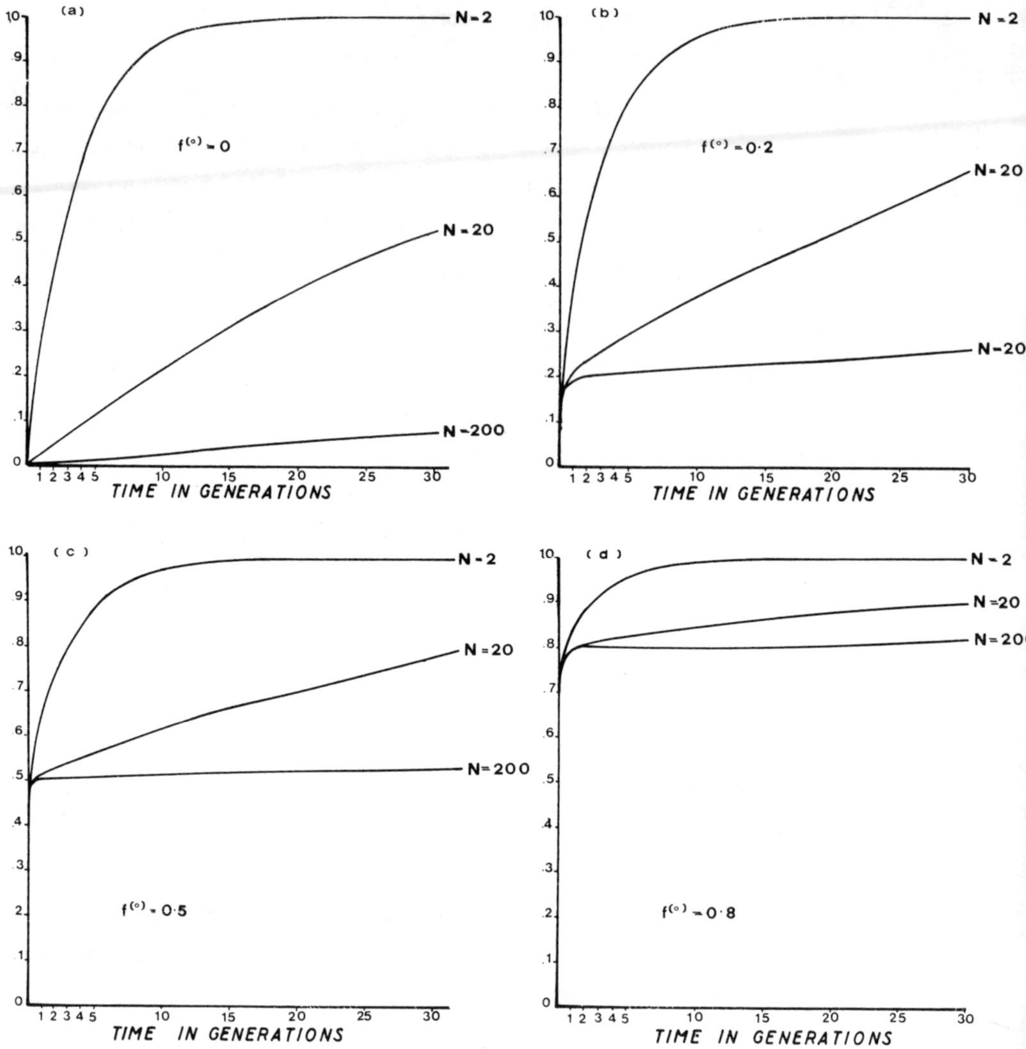

$$f_t = \frac{1}{2N} + \left(1 + \frac{1}{2N}\right) f_{t-1}$$

Fig. 6. Influence of the size of the population on its inbreeding coefficient through time when the initial inbreeding coefficient in the population is: (a) $f^{(o)} = 0$, (b) $f^{(o)} = 0.2$, (c) $f^{(o)} = 0.5$, (d) $f^{(o)} = 0.8$. A finite population ($N < 200$), (in the absence of mutations, selection etc.) leads not only to homozygosity but also to completely inbred individuals.

The importance and consequences of pollen flow are proportional to the outcrossing rate and selection pressure against the incoming allele.

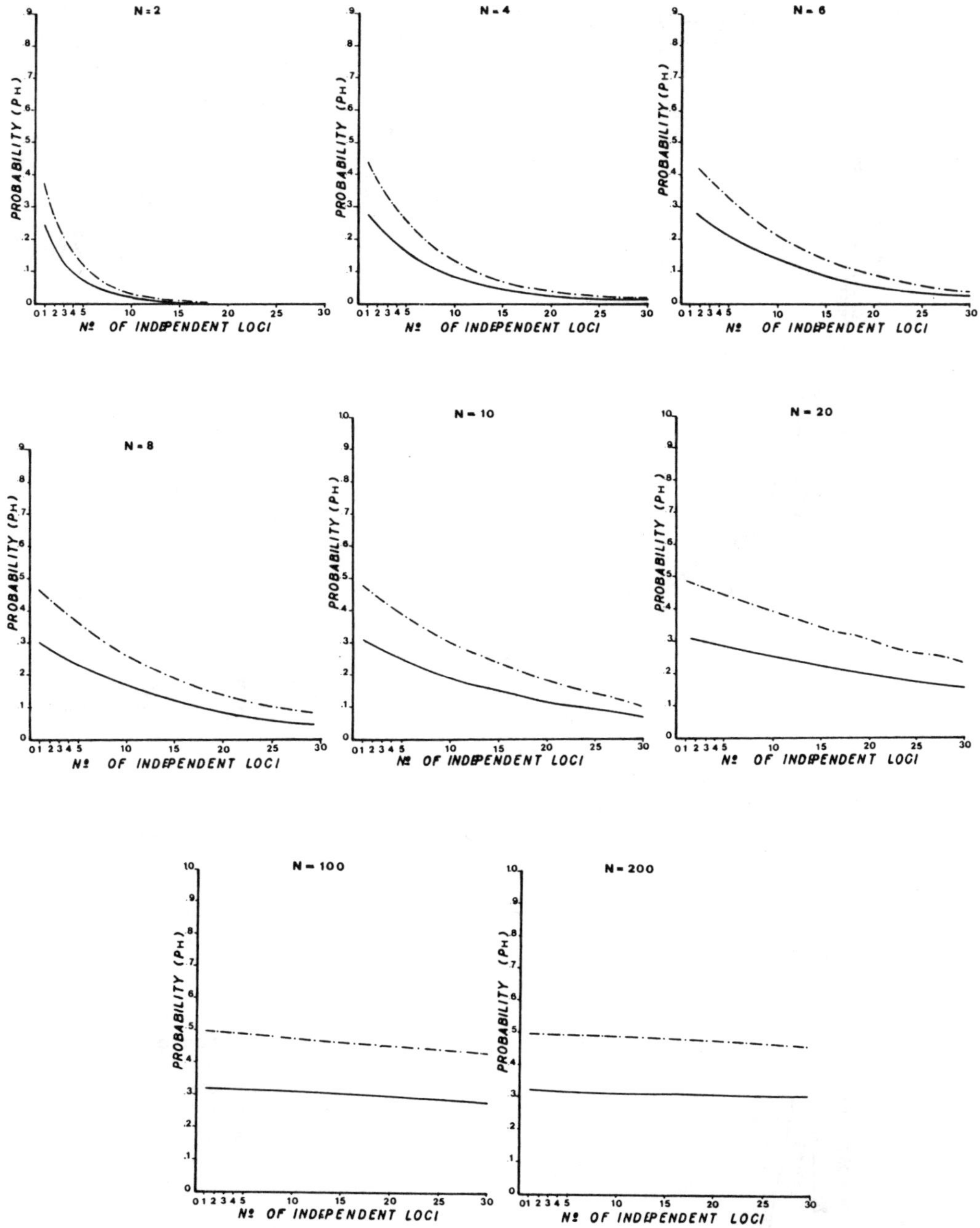

Fig. 7. (a) Probability (PH) that a randomly chosen individual in a finite sized population at the t-th generation is heterozygous. (b) Equivalently, if we consider many unlinked loci, the graph shows the probability that an individual is heterozygous at a fraction p of its loci. Equation Ht = 2 pq (1 - 1/2N)t was run for values of N equal to 2-4-6-8-10-20-100-200- and with initial A allele frequency of p equal to 0.2, 0.5 and 0.8.

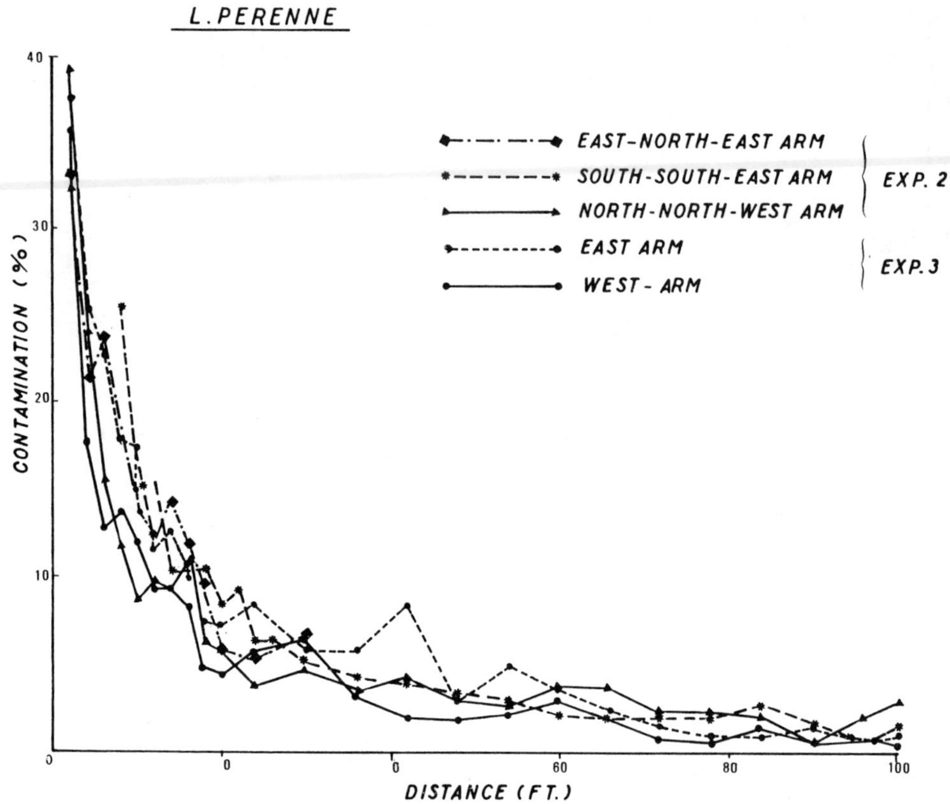

Fig. 8.a. Effect of isolation distance on the relative decrease in contamination (i.e. pollen dispersal) in *Lolium perenne* (Griffiths, 1950).

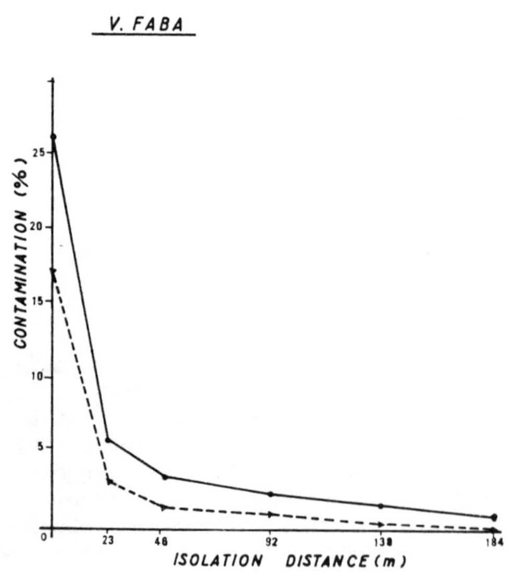

Fig. 8.b.
Contamination within the experimental field in relation to isolation distance •——• coloured-flowered plants from crosses with S43 as percentage of number established ◆——◆, coloured-flowered plants from crosses with S43 as percentage of number sown. (Bond and Pope, 1975).

If gene flow occurs, the quantity which characterises its main effects is the root-mean-square migration length, (l^2) whose expression is:

$$l^2 = \int x^2 \; g(x) \; dx$$

where g(x) is a function which gives the probability for an individual to move a distance $\underline{x}$ in unit of time, usually in one generation. Slatkin (1973) showed that a population cannot respond to changes in environmental condition which occur over a distance of less than the characteristic length. The result does not depend either on the pattern of variation in selective pressures or on the exact shape of the dispersal function g(x). The reduction in the fitness of the heterozygote causes the gene frequencies of two populations to become more differentiated.

A barrier to dispersal causes a drastic change in gene frequencies at the barrier only when almost all the gene flow is stopped and selection pressures continue to act in the population on the boundary.

The first experimental evidences of the effect of spatial isolation on pollen contamination have been given by Griffiths (1950) in *Lolium perenne* and later by Bond and Pope (1974) in *V. faba* (Figure 8).

When accessions, belonging to the same species, are grown at the same time in the same field, each sample may actually exchange pollen with other accessions. Gene flow could have no effect if it is bi-directional and the migrant allele has the same frequency and fitness in each population. Populations exchanging migrants at a constant rate (m) and in absence of differential allele fitness, converge to a common equilibrium frequency (Figure 9).

Antonovics (1968) studied the effect of gene flow, using computer simulation. When the pollen flow is uni-directional and concerns only the pollen from genotypes containing aa, then the $\underline{a}$ allele rapidly spreads through the population, unless selection acts against it; in this latter case the result will depend on the selection pressure, the rate of gene flow and the degree of dominance. Less important is the initial gene frequency. Pollen flow is effective in maintaining polymorphism in spite of strong selection against the incoming allele. For

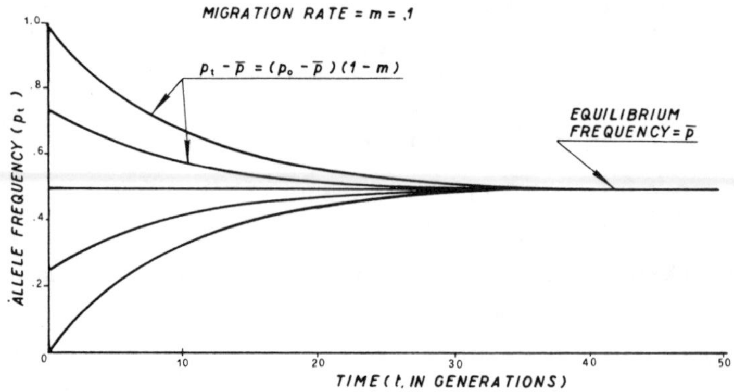

Fig. 9. Change of allele frequency against time in five sub-populations exchanging migrants at the rate m = 0.1/generation. Note rapid convergence to a common equilibrium frequency. (From principles of population genetics by D.L. Hartl, Sinauer Ass., Inc. Publ., Massachusetts).

example, pollen flow implying the a allele at a rate of 0.1 will reduce the frequency of favoured A allele from 1.0 to 0.68 if A is dominant and in spite of selection pressure against the incoming allele a equal to 0.4 (Figure 10). Intense pollen flow can have serious consequences on germplasm multiplication. The frequency reached by the favoured allele depends on the selection pressure against the incoming allele irrespective of initial gene frequency. The highest changes in the frequency of the favoured allele occur in the first ten generations of multiplication if selection coefficients are of the order of 0.2 and 0.4 against the incoming allele (Figure 11).

When pollen flow regards only a favoured allele, say A, then under relatively strong selection pressures the favoured A allele increases to a frequency approaching its final equilibrium value within the first 5 to 20 generations (Figure 12).

Jain and Bradshaw (1966) showed that complete dominance and unequal generation times tend to give slightly lower values of the A allele frequency at equilibrium but rates of approach to equilibrium are high. Thus, a favourable gene with such a distinct selective advantage could easily be increased in the population over a very few generations, irrespective of the

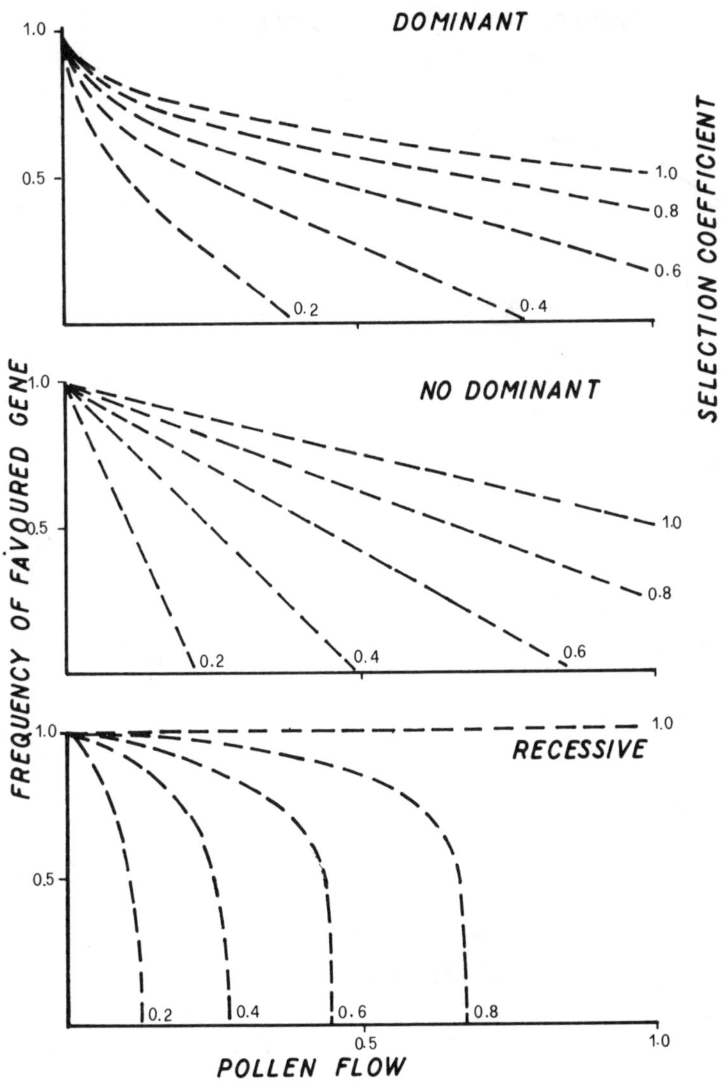

Fig. 10. Effect of selection and pollen flow on gene frequency and heterozygosity at equilibrium. (Selection coefficients, and degree of dominance of favoured gene, are indicated on graphs). (Antonovics, 1968).

form of migration. Even with 50 - 60% pollen contamination, at any site, selection pressure equal to 0.3 or higher is able to maintain local differentiation. Under weak selection, gene flow results in the fixation of the most frequent 'dispersed' allele. At any rate, the joint relationship between the opposing forces of selection and migration tends to become very complex.

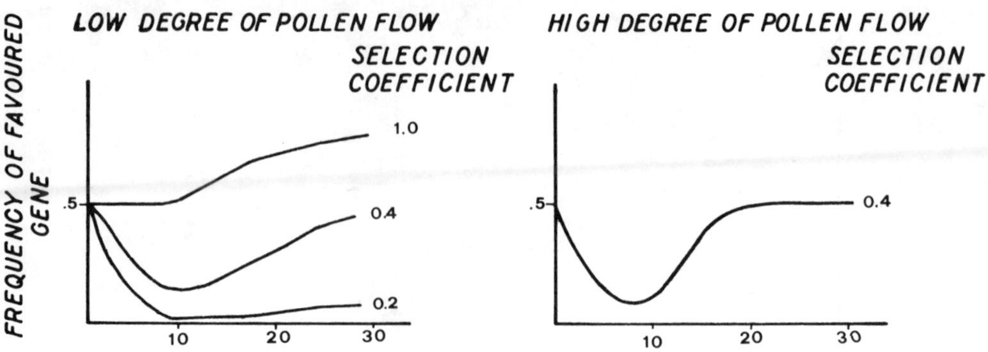

Fig. 11. Effect of changing pollen flow on gene frequency - favoured gene is dominant. (Adapted from Antonovics, 1968).

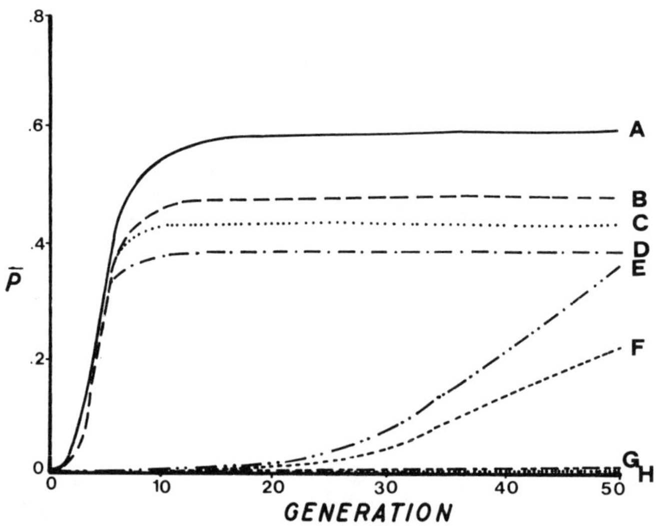

A = S1 = 0.95,	S2 = 0.50;	No dominance	T1 = T2 = 1
B = S1 = 0.95,	S2 = 0.50;	" "	T1 = 1, T2 = 5
C = S1 = 0.95,	S2 = 0.50;	Complete dominance	T1 = T2 = 1
D = S1 = 0.95,	S2 = 0.50;	" "	T1 = 1, T2 = 5
E = S1 = 0.25,	S2 = 0.10;	" "	T1 = T2 = 1
F = S1 = 0.25,	S2 = 0.10;	" "	T1 = 1, T2 = 5
G = S1 = 0.25,	S2 = 0.10;	No dominance	T1 = T2 = 1
H = S1 = 0.25,	S2 = 0.10;	" "	T1 = 1, T2 = 5

Fig. 12. Rates of initial increase of allele A under dispersal model 4A. (Jain and Bradshaw, 1966).

Two loci theory

A good agreement is generally found, between theoretical principles and experimental results, when a single locus is considered. The situation is quite different when two or more loci are analysed (Karlin, 1975; Karlin and Carmelli, 1975a; b).

In the latter case, new factors and some complexity such as gene recombination and broad sense epistasis need to be added to the interallic effects, such as additivity and dominance, typical of the single locus situation.

One of the essential concepts in a two loci situation is the linkage equilibrium (Geiringer, 1944), describing the condition in which gametic frequencies are the product of the appropriate allele frequencies.

Gametes	FREQUENCY At equilibrium	At non-equilibrium
AB	$\hat{g}_{11} = pr$	$g_{11} = \hat{g}_{11} + D$
Ab	$\hat{g}_{10} = p(1-r)$	$g_{10} = \hat{g}_{10} - D$
aB	$\hat{g}_{01} = (1-p)r$	$g_{01} = \hat{g}_{01} - D$
ab	$\hat{g}_{00} = (1-p)\ (1-r)$	$g_{00} = \hat{g}_{00} + D$

where p is the frequency of allele A, (1-p) the frequency of allele a, r the frequency of allele B and (1-r) the frequency of allele b.

At non-equilibrium, genetic frequencies are loaded by a quantity D, ($D = \hat{g}_{11}\ \hat{g}_{00} - \hat{g}_{10}\ \hat{g}_{01}$), which measures the statistical dependence between two loci. At equilibrium D = 0.

When two loci are independent in their functions, for example one controls flower colour and the other plant height, one locus theory can be properly used for each of them. But when two loci are simultaneously involved in the expression of a character, selection acts on both of them, causing the aggregation of 'packets' or 'complexes' of alleles.

In this condition linkage and epistasis produce gene and/ or genotypic frequencies that, at equilibrium, allow the highest fitness (Wright, 1942, 1949; Lewontin and Kojima, 1960).

Any change in one of the two factors will produce a change in the population structure and such a change will be stronger when marked epistasis and tight linkage occur simultaneously. In extreme situations any stable equilibrium cannot be reached and the population tends to fixation.

The amount of recombination allowed for such cases is a function of epistatic deviations. If linkage is tighter than the value demanded by the magnitude of epistatic deviations, there may be permanent disequilibrium of considerable magnitude, and gene frequencies may also be affected. Linkage among non-interacting genes does not build up disequilibrium of linkage.

In large populations epistasis in selective value of different genotypes is necessary to maintain linkage disequilibrium (Lewontin and Kojima, 1960). For finite unselected random mating populations, linkage disequilibrium could be assessed between segregating loci; expressions for changes in linkage disequilibrium have been given by Kimura (1963) and by Hill and Robertson (1966). Hill and Robertson (1968) advise caution in concluding that a significant departure from equilibrium, when observed in a small population, is due to natural selection. Any restriction to population size may cause disequilibrium as a result of genetic sampling, and the return to equilibrium will be slow if the loci are tightly linked.

Significant departure from equilibrium is also observed in small populations as a result of genetic sampling. Return to equilibrium is generally slow when the loci are tightly linked.

The effects of the mating system at two loci have been studied by several authors (Bennett and Binet, 1956; Kimura, 1958; Ghai, 1964; Narain, 1966; Karlin, 1969; Weir and Cockerham, 1973). Geiringer (1944) showed that for s (proportion of selfing) $\neq 1$ and λ (degree of linkage) $\neq 1$, original linkage blocks are finally broken up, while for $s = 1$ any one of the two non-allelic form of the genes in the zygote has a probability of $1/2 - \lambda$ of having been on the same gamete originally. Original gametes remain intact for $\lambda = 1$.

With strong selection against single heterozygotes, an increase in selfing determines a decrease in gametic disequilibrium (Holden, 1979); the change is proportional to the degree of linkage.

With much selfing, the frequency of heterozygotes can be so small that the decay and build-up rate of disequilibrium are also small. This implies that a higher number of generations may be required to reach such an equilibrium value of D than in random mating.

Many of these principles are based on the assumption of a symmetry of viability values. In the absence of such a symmetry, the achievement of a stable polymorphic equilibrium with a large selfing fraction is more difficult. Allard and Hansche (1964) and Jain (1968) found evidence that under mixed mating the pattern of gene dispersion and the role of linkage and interaction in producing gametic imbalance varied only in some quantitative details rather than qualitatively, compared to the situation when only random mating is involved or where only one locus is taken in consideration.

Jain (1968a), in a simulation study, showed that random drift for six linked loci in population of size N = 20 and for three levels of selfing (s = 0, s = 0.5, s = 0.9) the final situation at each locus is linked to the single locus case. However, fixation proceeds more rapidly with inbreeding but polymorphism at each locus is still present after five generations when selfing is 0.9 (Figure 13). Linkage reduces the dispersive effects of random drift where selection models favour situation of gametic imbalance.

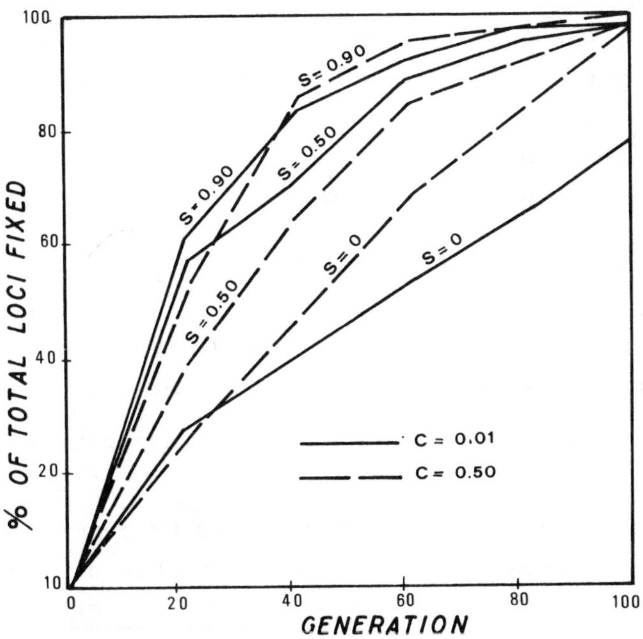

Fig. 13. Random drift for 6 linked loci in populations of size N = 20. All gene frequencies were initially 0.5 and no selection was assumed. The percentage of loci fixed was computed 20 runs for each pair of values of s (percent self-fertilisation) and c (recombination value) (Jain, 1968).

An increase in the number of loci determines a significant decrease in the rate of loss of heterozygosity and linkage affects heterozygosis (Fraser and Bumell, 1968). In general, for moderate to large population size, simulation studies on the loss of heterozygosity at several loci show results in line with those reported for two locus deterministic models.

In view of this evidence it is expected that two loci theory does not add any insight to the information gained from one locus theory. Therefore, the following considerations for a proper strategy of seed regeneration will be based mainly on the events associated with a diallelic locus for no more than ten generations.

Considerations and implications

The above mentioned basic aspects and research results allow the following considerations and implications.

Selection

The results of the analysis on the effects of selection, reported under heading 'Fitness', lead to some important considerations.

1) Generally, conditions which favour polymorphism favour the permanence of the alleles in the population as well.
2) When there is heterozygotic advantage and selection is the only force acting on the allele dynamics of a population, there are no problems in maintaining alleles in the population.
3) If selection against the dominant allele A is high, in three generations allele frequency will be reduced from 0.2 to 0.0. More generations are required to reduce the frequency of the recessive allele a, in the same conditions.

Effects of selection can be avoided by adopting appropriate techniques.

1) Variability in seed viability can be reduced by germinating seed in Petri dishes in order to get a number of seedlings proportional to the amplitude of each class of seed viability.

2) Competition for resources can be controlled by adopting a sowing density lower or equal to that in the usual agronomic practice of each field crop considered.

3) Selection, due to environmental factors such as frost, drought or diseases, can be avoided by reducing their strength through:

- anticipating or postponing sowing dates and/or physical protection;
- irrigation for avoiding differential survival due to water stress;
- using chemical control of insects and/or fungal parasites.

4) Variability in ripening time, which can result in differential loss of fitness because of dissemination *(Secale cereale)* and/or bird damage, can be avoided by harvesting each genotype or group of genotypes at the proper time or by protecting plots under a net. With a large variability in fitness, selection pressure can be avoided by collecting the same number of seeds from each genotype, for example one spike in case of rye *(Secale cereale)* or one pod in case of broad bean *(Vicia faba)*. With a narrow fitness variance, occurring when most of the genotypes have the same fitness while very few are different, selection can be avoided by dividing the sample in two or more sub-populations each of which contains genotypes with the same fitness. Each sub-population will be collected and threshed altogether without any other care. After threshing, sub-populations can be reassembled introducing correction factors for disadvantaged genotypes.

5) In species such as maize and sunflower, for which hand-crossing is possible all the different genotypes must contribute with the same amount of pollen to the pollen-pool used for pollination.

<u>Mating system</u>

In the absence of selection, gene frequencies remain constant through generations, irrespective of the mating system.

1) If selection forces are strong enough, the loss of

unfavoured alleles is speeded up by increased selfing. So the probability of losing genes can be reduced by increasing the outcrossing rate.

2) In mixed selfing and random mating, the effect of selection in modifying the allele frequencies is lower than in random mating. Then in cross-pollinated species, such as maize or sunflower, an increase of selfing should decrease the rate of change in allele frequencies. In *Vicia faba*, seeds from basal pods are mainly the result of outcrossing while seeds from apical pods are mainly due to self fertilisation (Porceddu et al., 1980). In this case seeds from four pods on the lower part and two pods from the top part of the plant should be collected and pooled together.

3) Assortative mating increases homozygote frequency which can be further increased if significant selection pressure is present. For example, this can be due to wide differences in flowering time existing among genotypes. This inconvenience can be overcome by sowing groups of plants at different times; ten to fifteen days between two sowings for summer and one month for winter crops.

Inbreeding

Inbreeding modifications can be evaluated, both:

- indirectly: by evaluating modifications of the outcrossing rate in the new environment;
- directly: by evaluating modifications in fitness variability (an increase of fitness variability can be associated with increasing inbreeding) or by testing the isoenzyme polymorphism or by analysing the frequency of morphological markers; knowledge of the initial inbreeding rate, agronomic techniques and management can be of help in this connection. For example peasants usually grow a very limited number of plants of maize and sunflower in their fields. It is expected that the populations that they grow reach the same high level of inbreeding.

The desired level of inbreeding can be kept constant by:

1) Modifying the mating system by artificially increasing or decreasing the selfing rate;

2) Choosing a proper population size.

An increase in the inbreeding rate results for two reasons:

- Some alleles can be lost because selection effects, against genotypes showing inbreeding depression, are stronger.
- The percentage of homozygous genotypes increases in the population as a result of sampling effects. In the latter case an adequate number of plants must be grown if all alleles need to be preserved. Maintaining heterozygosity is the most advantageous way to gene preservation. For this reason the minimum number of plants would become proportionally higher if all alleles are preserved and the heterozygous form is the most advantageous for gene preservation.

Population size

When population size is reduced inbreeding rate increases, favoured alleles tend to be fixed and unfavoured ones to be lost.

1) Since, in a random mating population, heterozygotes contribute both alleles, the probability of finding a heterozygous genotype in the population will give a measure of a chance of finding both alleles after a number of generations.
2) The minimum number of individuals, to be taken from a population to have a 40% chance of getting a heterozygote individual at a single locus, when only the dispersive process affects gene frequency, is three in the first generation, four in the second and from 10 to 20 after the fifth generation.
3) The population size must be increased at each regeneration, if heterozygous types have to be kept because of the increase of homozygous types due to sampling error.
4) Population size can be kept constant by favouring selection against homozygous types.
5) With a population composed of 100 plants, the probability of finding a heterozygote is higher than 0.4, even after 30 generations, if the initial frequency was 0.5. This figure gives an idea of how much the population structure can change as a consequence of its size through generations.

6) When multi-loci situations are considered the probability of finding thirty independent loci all in a heterozygous state, is higher than 0.4 in a finite population of size N = 100, and higher than 0.5 in infinite populations.

7) When the number of independent loci is lower, the difference between finite and infinite populations becomes even slighter. Linkage and inbreeding considerably reduce the number of independent loci, so that the number of plants to be grown at each generation can be between 20 and 200, according to the species and inbreeding rate. The higher the inbreeding level the lower should be the number of plants requested to represent the whole set of alleles. This limit should not be beyond the value requested to maintain the inbreeding at the level already reached.

Migration

Gene flow can cause different degrees of alteration of the genetic structure of populations depending on:

1) Whether populations share the same gene pool and possess almost the same frequencies, or

2) Whether populations are differentiated because of a different gene pool.

In the first case, gene flow results in a modification of the allelic frequency and if gene flow is not uni-directional, the allele will not easily be lost.

In the second case, greater damage to germplasm resources can result.

The damage depends upon the following situations:

1) When an unfavoured allele a having selective pressure against it equal to 0.4 migrates at a rate of 0.1, the percentage of the favoured allele A is reduced from 1 to 0.68 in one generation.

2) If the migrant allele A is the favoured one, ten generations are necessary to lower the frequency of the unfavoured allele a from 0.5 to 0.2.

3) If the migrant allele A has only a slight advantage over the allele a its presence in the population increases from 0.001 to 0.4 in five generations.

4) It is necessary to have a disadvantage of 70% against the migrant allele to keep the population differentiated in the presence of a pollen flow of 50 - 60%.
5) Gene flow can be completely counterbalanced by selection forces. Selection can maintain population identity even if samples are grown at very close distance.
6) The general pattern of decreasing gene flow-rate over distance varies little with a variety of species, both insect and wind pollinated.

Gene flow can be due to seed or pollen flow. Seed flow can be easily prevented by using an accurate sowing technique and self-cleaning threshers easily inspectable. Pollen flow can be prevented by means of:

a) spatial isolation, although it is often difficult to achieve while growing germplasm collections;
b) artificial barriers such as nets, or natural barriers such as plants of a taller species surrounding each population plot (reduced insects activity among plots will result);
c) hand-crossing anemophylous crops like rye and maize, or the entomophylous like sunflower;
d) collecting seeds from the part of the plant where it is supposed to have a lower degree of out-crossing as in the apical pods of *V. faba*;
e) artificial selection against the migrant allele can reduce the effect of gene flow. This implies easy recognition of the migrant allele in genotypes carrying it.

So for practical purposes, if gene flow cannot be prevented the possibility of growing, in the same field, populations sharing approximately the same gene pool or coming from the same environments or from the same geographical area must be considered.

EXPERIMENTAL ANALYSIS OF FACTORS AFFECTING POPULATION STRUCTURE IN SOME FIELD SPECIES

The genetic models and their implications in maintaining genetic variability, presented in the previous section, evidenced as the parameters that are relevant in determining population structure over time are those describing the mating system and relative viabilities in the case of a one locus genetic system. For genetic systems involving two or more loci, additional parameters describing recombination rate and epistatic effects between loci are needed.

Several data are now available for some field crop species which allow either a direct or indirect estimation of factors that will explain the genetic ratios observed in natural and experimental populations.

In this section the quantitative estimates of the parameters obtained from actual population of four field crop species and some of the indications that such estimates give for developing a proper strategy for maintaining the original genetic variability of the sampled populations during seed regeneration will be considered.

The four crops, maize, rye, sunflower and broad beans were chosen as an example due to the peculiarities of their breeding systems (Table 1).

In *Helianthus annuus* each flower head, or calatide, is composed of a large number of florets (between 1 000 - 2 000 florets/flower-head).

In field condition it is largely dependent on bees both for self-and cross-pollination because of the floral structure.

Only one percent of florets/flower-head get fertilised if insects are absent (Free and Simpson, 1964). Generally sunflower is self-fertile but varietal differences, and reduction of self-fertility attributed to environmental factors (Alex, 1957a; Pinthus, 1959) have been reported.

Out-crossing rate ranges from 50 - 78% and cross-pollination is normally favoured (Putt, 1940; Furgala, 1954). Cross-pollination readily occurs and a distance of 1.6 km between sunflower seed crop fields should be maintained if a very low rate of genetic contamination is desired. At shorter distance, pollen flow from other sunflower fields is heavy.

Alba (1981) (unpublished data) noticed that cross-pollination decreases from 60 to 40% if the distance between plants carrying a dominant marker gene and others without the marker gene is increased from 60 cm to 180 cm.

TABLE 1

PRACTICAL ASPECTS

	Zea mays	*Helianthus annuus*	*Vicia faba*	*Secale cereale*
Population size	20-100 plants well spaced	20-100 plants well spaced	20-80 plants well spaced	100-200 plants well spaced
Pollen flow prevention	Use paper bags to cover randomly chosen spikes	Use bags to isolate randomly chosen inflorescences	100 m of distance between plots or less if artificial or natural (plant) barriers are present	Bag together at least 100 spikes
Mating system	Artificial inter-crossing 8-20 female plants with pollen pool from all the others	Artificial inter-crossing 8-20 female plants with pollen pool from all the others	Allow insects pollination	Natural inter-crossing among neighbouring plants
Harvesting	Same number of seeds 150 from each spike	Same number of seeds 100 from each inflorescence	Same number of seeds 10 from each plant; half from apical, half from basal nodes	Same number of seeds at least 9 from each plant

Investigations on the minimum number of plants necessary to stabilise the genetic characteristics of a variety have not been experimentally determined. A test carried out with two varieties, aimed at establishing the number of plants to maintain a high open pollinated variety through generation, suggested that limitation on the number of plants between 25 and 100 does not modify consistently genetic diversity of the population. If assortative mating is allowed stable allele frequencies at a locus are expected. (To do this, isolate single calatides of the same progeny with paper bag, sib-pollinate them, harvest four pollinated calatides and two sib-pollinated calatides belonging to the same progeny; mix the seeds and grow 100 plants from those seeds).

The population size and an appropriate mating system are the same as when dealing with maize. Flowers can be isolated with cloth bags. Assortative mating, due to differences in flowering time, can be largely avoided by repeatedly pollinating the same inflorescence, because of the different flowering time from the centre to the periphery of the calatides.

Zea mays is a highly cross-pollinated species and, even if compatible, self-pollination never occurs at a rate higher than 10%. Self-pollination can be sometimes prevented by protogyny or by protandry because of a too wide difference in time of maturity of the two inflorescences (Weatherwax, 1955).

The number of pollen grains produced from each plant is incredibly high. Experiments with commercial corn indicate that the proportion of pollen grains producing kernels is around 1 : 50 000 (Weatherwax, 1955). Commercial seed fields are kept 200 m apart to avoid contamination due to pollen flow from other varieties but the choice of this distance is based 'on practical experience rather than on experimental evidence' (Gacitua, 1946).

The greatest amount of contamination occurs, nevertheless, within the first 50 - 75 m of distance among populations. Barriers can consistently reduce this distance. Direction of contaminating corn or 'depth of field' in the direction of contaminating pollen influences the amount of contamination (Gacitua, 1946; Airy, 1950; Jones and Brooks, 1950 - 52). Genetic drift has probably played a major role in the evolution of local populations.

Experiments reported by Sprague (1939) have shown that, to represent adequately the genetic variation in a North American corn variety, ten to twenty plants are required.

Indian maize growers from North America probably saved few spikes to obtain seeds to be sown in their small fields and sometimes new populations have been produced from a single spike (Mangelsdorf, 1974). Evidence of genetic drift resulting in higher level of inbreeding has been reported in many regions of Asia. A relatively high frequency of waxy endosperm type, due to a recessive simple inherited character, has been reported by several authors (Collins, 1909, 1920; Kuleshov, 1928; Stonor and Andersen, 1949) and attributed to inbreeding. On the other hand artificial selection has been largely performed, resulting in reduced genetic variability of maize populations (Wallace and

Brown, 1956; Anderson, 1947; Wellhausen et al., 1957; Mangelsdorf, 1974).

The population size should range from 20 (Sprague, 1939) to 100 (Oka, 1969) plants depending on inbreeding rate and the variation of characters related with the fitness. A proper mating scheme would be the artificial inter-crossing of 7 - 8 plants chosen as female parents from the different classes of flowering time in the population and isolated with paper bags. A pollen mixture must be obtained from 20 to 100 plants assuming that each genotype contributes the same amount of pollen. This pollen mixture should be used to hand pollinate two to three times the female plants. The same number of seedlings from the seeds of each spike should be raised to give to all the plant genotypes the same probability to reach the next generation.

Secale cereale. Because of the high self-sterility level, out-crossing is obligate and the inbreeding rate almost nil. When cross-pollination is prevented 96 to 99% of the spikelets of a spike have empty florets. Pollen flow is around 0.5 when populations are grown adjacent and reduces to 0.1 when distance is increased to 3 m (Sprague, 1938). In this case it is not so difficult to maintain genetic variability of several loci by growing a number of plants for each population ranging from 80 to 120 depending on fitness variation and leaving 5 - 10 m between plants belonging to different rye populations. The best technique, however, will be to cage the plants belonging to the same population and provide a means for pollen flow within the cage. The same method can be adopted with maize and sunflower taking care to multiply by ten the number of female parents and by two the male parents. Another possibility is to bag together with paper bags five to seven spikes from different plants: one in the centre and four to six all around.

The third possibility, which is less expensive and also less reliable, is to grow all around each plot two to four rows of sunflower or corn in order to impede the dispersal of the pollen over different population plots. Finally one spike with about the same number of seeds from each plant should be collected at harvest time.

Vicia faba. Pope and Bond (1975) showed in *V. faba* that the genetic contamination decreased from 16.96% at 0.9 m to 1.25%

at 92 m and 0.59% at 184 m. Maximum contamination from other *V. faba* plants outside the experimental field and at least 1.1 km away was 0.44%

Populations grown in plots measuring 4.5 m x 4.5 m, containing at least 40 plants could show a contamination of between 1 and 2% if spaced at 92 m. This is unlikely to affect their genetic identity if selection intensity (either natural or artificial) is maintained against the incoming allele. Such selection intensity should be around 0.2 in order to counterbalance the pollen flow effectively.

A consistent level of inbreeding must be taken into account in this species because of the low out-crossing rate, ranging from 25 to 50%. The optimum population size can be in this case from 20 - 28 plants. Because of the need of insects as pollinators three different strategies can be suggested.

a) to grow each population in a separate net cage and introduce bees at flowering time;

b) to grow a number of plants at least two times larger than that necessary for regeneration purposes and to collect seeds only from the plants in the central part of each plot;

c) to grow in the same field with the same technique as before (b) only populations from the same homogeneous geographical area. A distance of at least 200 m is required between different multiplication fields.
Two pods with the same amount of seeds should be collected from each plant: one from the basal and the other from the apical nodes.

ACKNOWLEDGEMENTS

The authors wish to thank Mrs. M.G. De Luca (Germplasm Institute, CNR, Bari) for typing the original manuscript.

REFERENCES

Airy, J.M., 1950. Current problems of detasselling. Rept. Fifth Ann. Hybrid Corn Industry Research Conf., pp. 6-10.

Alex, A.H., 1957. Pollination of some oilseed crops by honeybees. Progr. Rep. Texas Agric. Exp. Stn., No. 1960.

Allard, R.W. and Hansche, P.E., 1964. Some parameters of population variability and their implications in plant breeding. Advan. Agron., 16: 281-325.

Anderson, E., 1947. Field studies on Guatemalan maize. Ann. Missouri Bot. Gard., 34: 433-467.
Antonovics, J., 1968. Evolution in closely adjacent plant populations. VI. Manifold effects of gene flow. Heredity, 23: 507-524.
Bennett, J.H. and Binet, F.E., 1956. Association between Mendelian factors with mixed selfing and random mating. Heredity, 10: 51-55.
Bond, D.A. and Pope, M., 1974. Factors affecting the proportions of cross-breed and selfed seed obtained from field bean (*Vicia faba* L.) crops. Journal of Agricultural Science, Cambridge 83: 343-351.
Collins, G.N., 1909. A new type of Indian corn from China. USDA Bur. Pl. Ind. Bull., 161.
Collins, G.N., 1920. Waxy maize from upper Burma. Science 52: 48-51.
East, E.M., 1916. Inheritance in crosses between *Nicotiana Langsdari* and *Nicotiana Alata*. Genetics, 1: 311-333.
Fraser, A.S. and Burnell, D.G., 1968. Computer genetics. New York: McGraw Hill Pub.
Free, J.B. and Simpson, J., 1964. The pollination requirements of sunflowers (*Helianthus annuus* L.) Emp. J. Exp. Agric., 32: 340-342.
Furgala, B., 1954. The effect of the honey bee *Apis mellifera* (L.) on the seed set and hybridization of the cultivated sunflower *Helianthus annuus*. Ph. D. Thesis Univ. Manitoba.
Gacitua, H.L., 1946. Factors which determine the importance of contamination in corn. MS Thesis Iowa State College, Ames, Iowa.
Geiringer, H., 1944. On the probability theory of linkage in Mendelian heredity. Annals of Mathematical Statistics, 15: 25-57.
Ghai, G.L., 1964. The genotypic composition and variability in plant populations under mixed self-fertilization and random mating. Jour. of the Indian Society of Agricultural Statistics, 16: 94-125.
Griffiths, D.J., 1950. The liability of seed crops of perennial rye grass (*Lolium perenne*) to contamination by wind-borne pollen. J. Agric. Science, 40: 19-38.
Hayman, B.I., 1953. Mixed selfing and random mating when homozygotes are at a disadvantage. Heredity, 7: 185-192.
Hill, W.G. and Robertson, A., 1966. The effect of linkage on limits to artificial selection. Gen. Res. 8: 269-294.
Hill, W.G. and Robertson, A., 1968. Linkage disequilibrium in finite populations. Theor. Applied Genetics, 38: 226-231.
Holden, L.R., 1970. New properties of the two locus partial selfing model with selection. Genetics, 93: 217-236.
Hurt, E.F., 1946. Sunflower. Faber and Faber London.
Jain, S.K., 1968. Simulation of models involving mixed selfing and random mating. II. Effect of selection and linkage in finite populations. Theor. and Appl. Gen., 38: 232-242.
Jain, S.K. and Bradshaw, A.D., 1966. Evolutionary divergence among adjacent plant populations. I) The evidence and its theoretical analysis. Heredity, 21: 407-441.
Jain, S.K. and Marshall, D.R., 1968. Simulation of models involving mixed selfing and random mating. I. Stochastic variation in outcrossing and selection parameters. Heredity, 23: 411-432.
Jones, D.F. and Brooks, J.S., 1950. Effectiveness of distance and border rows in preventing outcrossing in corn. Oklahoma Agr. Expt. Sta. Tech. Bull. T-38.
Jones, D.F. and Brooks, J.S., 1952. Effect of tree barriers on outcrossing in corn. Oklahoma Agr. Sta. Tech. Bull. T-45.
Karlin, S., 1969. Equilibrium behaviour of population genetic models with non-random mating. New York: Gordon and Breach.
Karlin, S., 1975. General two locus selection models. Some objectives results and interpretations. Theor. Pop. Biol., 7: 364-398.

Karlin, S. and Carmelli, D., 1975a. Numerical studies on two-loci selection model with general viabilities. Theor. Pop. Biol., 7: 397-421.

Karlin, S. and Carmelli, D., 1975b. Some population genetic models combining artificial and natural selection pressures. II. Two locus theory. Theor. Pop. Biol., 7: 123-148.

Kimura, M., 1958. Zygotic frequencies in a partially self-fertilizing population. Annual Report of the National Institute of Genetics, Japan, 8: 104-105.

Kimura, M., 1963. A probability method for treating inbreeding systems especially with linked genes. Biometrics, 19: 1-17.

Kuleshow, N.N., 1928. Some peculiarities in the maize of Asia. Bull. Appl. Bot. and Pl. Breed., 19: 325-374.

Law, C.N., 1967. The location of genetic factors controlling a number of quantitative characters in wheat. Genetics, 56: 445-461.

Li, C.C., 1976. First course in population genetics. Boxwood, Pacific Grove, California.

Lewontin, R.C. and Kojima, K., 1960. The evolutionary dynamics of complex polymorphisms. Evolution, 14: 458-472.

Mangelsdorf, P.C., 1974. Corn: its origin evolution and improvement. The Belknap Press of Harvard University Press, Cambridge, Massachusetts.

Narain, P., 1966. Effect of linkage on homozygosity of a population under mixed selfing and random mating. Genetics, 54: 303-314.

Oka, H.I., 1969. A note on the design of germplasm preservation work in grain crops. SABRAO Newsletter, 1: 127-134.

Payne, F., 1918. An experiment to test the nature of the variations on which selection acts. Indiana Univ. Stud., 5: 1-45.

Pinthus, M.J., 1959. Seed set of self-fertilized sunflower heads. Agron. J., 51: 626.

Pope, M. and Bond, D.A., 1975. Influence of isolation distance on genetic contamination of field bean (*Vicia faba* L.) seed produced in small plots. Journal Agric. Sci., Cambridge, 85: 509-513.

Porceddu, E., Monti, L.M. Frusciante, L. and Volpe, N., 1980. Analysis of cross-pollination in *Vicia faba* L.. Z. Pflanzen., 84: 313-322.

Powers, L., 1963. The partitioning method of genetic analysis and some aspects of its application to plant breeding. In "Statistical Genetics and Plant Breeding". Ed. Hanson, W.D. and Robinson, H.F. pp. 280-318. National Academy of Sciences, Washington.

Putt, E.D., 1940. Observations on morphological characters and flowering processes in the sunflower (*Helianthus annuus* L.). Scient. Agric. 21: 167-179.

Qualset, C.O., 1979. Mendelian genetics of quantitative characters with reference to adaptation and breeding in wheat. In "Proc. Fifth Int. Wheat Gen. Symp." New Delhi 1978, Vol. 2, pp. 577-590.

Sax, K., 1923. The association of size differences with seed-coat pattern and pigmentation in *Phaseolus vulgaris*. Genetics, 8: 552-560.

Slatkin, M., 1973. Gene flow and selection in a cline. Genetics, 75: 733-756.

Sprague, H.B., 1938. Breeding rye by continuous selection. J. Am. Soc. Agron., 30: 287-293.

Sprague, G.F., 1939. An estimation of the number of top-crossed plants required for adequate representation of a corn variety. J. Am. Soc. Agron., 31: 11-16.

Stewart, J., 1969. Biometrical genetics with one or two loci. I. The choice of a specific genetic model. Heredity, 24: 211-224.

Stonor, C.R. and Andersen, E., 1949. Maize among the hill people of Assuan. Ann. Missouri Bot. Gard., 36: 355-404.

Wallace, H. and Brown, W.L., 1956. Corn and its early fathers East Lausing. Michigan State University Press.

Weatherwax, P., 1955. Corn and corn improvement. G.F. Sprague Ed. Academic Press Inc. Publishers, New York, 89-121.
Weber, E., 1959. The genetical analysis of characters with continuous variability on a Mendelian basis. I. Monohybrid segregation. Genetics, 44: 1131-1139.
Wehrhahan, C. and Allard, R.W., 1965. The detection and measurement of the effects of individual genes involved in the inheritance of a quantitative character in wheat. Genetics, 51: 109-119.
Weir, B.S. and Cockerham, C.C., 1973. Mixed self and random mating at two loci. Genet. Res. Camb., 21: 247-262.
Wellhausen, E.J., Roberts, L.M. and Hernandez, E., 1957. In collaboration with P.C. Mangelsdorf 1957. Races of maize in Central America. Nat. Acad. Sci. - Nat. Res. Council Publ. 511.
Wright, S., 1942. Statistical genetics and evolution. Bull. Am. Math. Soc., 48: 223-246.
Wright, S., 1949. Adaptation and selection. In "Genetics, Paleontology and Evolution" Ed. Jepan, G.L., pp. 365-389. Princeton University Press. New York.

DISCUSSION

J.G. Hawkes *(UK)* Your presentation has been a very useful guide to procedures for the future. I wonder if you would like to comment on whether the five examples you showed in the last slide could be used for models for other plants. Are these five models into which we can group other field crops, or must each one be considered on its own merits?

C. de Pace *(Italy)* I think I could say yes, with one reservation. The amount of outcrossing was in the range of 20% to 80% - not higher than 80% and not lower than 20%. In this case, the scheme you are suggesting would be extended to other field crop populations.

E. Porceddu *(Italy)* I would like to add a point here. The four crops were chosen for their particular mechanisms of pollination, their ways of bearing seeds, pods, spikes or heads, different paternity at different seasons of pollination etc. So a number of crops could be grouped - could be assimilated, one of those could be grouped between two of the models.

J.G. Hawkes It would be very useful for the publication which will follow this meeting to add a few more indications on the basis of this paper, if you felt you dared do that.

E. Porceddu Well, in fact, we were obliged to restrict the presentation to a few examples from all the data we have, because of time restrictions.

Y. Demarly *(France)* One of your conclusions was that it is better to have a special plant to reduce competition. This is probably true bearing in mind the assumption you made at the beginning. However, with some plants such as lucerne, if you want to keep the integrity of a population, you should not avoid some competition because it is an essential parameter of the population. When I was a lucerne breeder the best cultivar was the variety Du Puits, my idea was to look at the competition of the cultivar and I observed that most of the yield came from 10% of the population. The normal procedure was to suppress all the weak plants in order to have a better variety. By doing this I got a new variety which was lower than the first one because the weak plants were kind of 'slaves' to the high yielding plants; they helped them at the young stage and they disappeared afterwards. So, although they do not play a part in the pollination, as tetraploid and very heterzygous plants, at each generation new weak plants appear and play their part in the results of the high yielding plants. So, if you reproduce these kinds of populations without competition you allow the weak plants to participate in seed production and you will disturb the normal strategy of the cultivar.

C. de Pace I don't know very much about lucerne but in field crop plants it seems necessary to maintain a very low competition level. One of the theoretical aspects of competition to be considered is that on every occasion that there have been some genetical investigations, it was noticed that it has an effect in changing the population structure because competition affects, almost directly, the frequency of each genotype, which means lowering its contribution to the next generation. Maybe lucerne is a special case but generally speaking I would say that because it affects frequency, if it is not lowered, competition will also affect a certain level of selection pressure in the population and thus change the genetic composition of a population.

Y. Demarly I am not sure that lucerne is a special case because, for example, in some parts of Africa you have sorghum which is not sown but planted, and in this sorghum, if you suppress competition the cultivars are much more heterozygous than the sown cultivars. So, to preserve this kind of sorghum, you must have the same thing. If you suppress plantlets of sugar beet at the young stage you have the same problem of competition between seed plants.

C. de Pace At this point more experiments are required to find out what is the real role of competition in changing the genetic structure of the populations. Maybe the theory could be changed if more evidence was acquired from practical experiments.

Y. Demarly My comments were not a criticism of your work; I just wanted to draw your attention to the fact that it is not always the case.

E.L. Breese *(UK)* I think Professor Demarly underlines the dilemma that we face so specifically in these crops, almost all of them but particularly the perennial or the forage crops. The problem is the solution. If we talk about a population that is only maintained under high selection pressure, how do we preserve that? It seems, as a first idea, that we can only preserve it by maintaining that selection pressure. The problem is that we can never quite guarantee simulating that precise kind of selection pressure because there is not only competition. Competition exacerbates a particular management. That is why I think there is perhaps one way in which we can seek to secure the balance of the vegetative population and that is ultimately vegetative propagation, by whatever means, tissue culture, etc. The next best, and here is the debate and the argument, is where we free it from all kinds of selection pressure, including competition. In Aberystwyth at the moment we say we cannot emulate the selection pressures that preserve the population so we will do the next best, relying on a high heritability of the plant types there, to maintain the population through seeds with the minimum of selection pressures, including competition.

J.G. Hawkes In this connection I think it would be interesting to mention some comments of Sir Otto Frankel relating, in particular, to the regeneration of samples which are being regenerated away from the area in which they were collected, in other words, not in isoclimatic regions. Frankel's suggestion, which I think would be worth considering in this context, is that a great deal of selection pressure is exerted on materials, between individuals in the population, where the particular climatic differences will show themselves. In other words, if we are trying to regenerate in a climate very different from that where the material was grown, the selection pressures will be different and some individuals bearing certain alleles will disappear and others will not. One way of trying to solve this problem is to reduce individual competition. Frankel has suggested that with a cereal, for instance, whereas in normal sowing one has the plants very close together, that one should separate the plants so that each one is able to grow and yield on its own account without much, or perhaps without any, competition from its sister plants. One would then be able to harvest a yield from each plant separately and take the same number of seeds. Now, this is a matter of debate and it brings in the same sort of arguments and difficulties that we have been discussing. However, I think it would be worth looking at this suggestion to decide whether or not it might be relevant to the present discussion.

P. Spagnoletti *(Italy)* We have made a question at the very beginning. What we would like to know is if a regional population was sampled correctly. In this case we want to preserve in the population the whole structure, gene and genotype frequencies at the same time. But, I think it is very difficult to say that what we have sampled is a representative sample of the regional population. That is why we started by assuming that what we are multiplying is a sample of seed, and this is a new population. I don't know if it is too much to suggest this but I think in many cases you have a sample of seed that can be identified as a new population and this is the population structure that we should reproduce in a new environment in which there are selection pressures that we don't know. I don't know if this point was clear at the very beginning of the work.

Y. Demarly Dr. de Pace, was all your theoretical data made on the assumption of no epistasy?

C. de Pace We took epistatic effect into account through the evaluation of fitness.

H. Lamberts *(Netherlands)* Has anyone any comments on what Professor Hawkes has said in relation to Sir Otto Frankel?

E.L. Breese I had the feeling that what Professor Hawkes said about Sir Otto Frankel's ideas were, in some sense, commensurate with the ideas I was outlining earlier and might be complementary to them in that context. Perhaps we could discuss the absolute details later.

Studies on maintenance of rye (*Secale cereale* L.s.l.) collection

B.MOLSKI, M.MAŁUSZYŃSKA, R.KUBICZEK & W.ŁUCZAK
Botanical Garden of the Polish Academy of Sciences, Warsaw

INTRODUCTION

The Botanical Garden of the Polish Academy of Sciences in Warsaw carries the collection of *Secale* covering 1 194 accessions including 12 species and 843 cultivars.

Rye collection regeneration and multiplication requires isolation of each plot during flowering to avoid cross-pollination. The main aim of these studies is to define the conditions for propagation of rye in a field collection. Experiments on the influence of different methods of sowing and methods of isolation of field plots on the phenology and yield structure were carried out over six years in two locations.

Plots of 1.5 m^2 were used, since an area of this size can accommodate between 150 and 400 plants. In such heterogeneous taxons as *Secale* an optimal number (150) of plants can be grown as one intercrossing population, ensuring preservation of the gene pool.

MATERIALS AND METHODS

A three year experiment, 1974 - 1976, was conducted in Skierniewice on light clay soil. During the next three years of study, 1977 - 1979, the experimental fields were located in Powsin on light argillaceous sand. The winter rye variety, Dańkowskie Złote, was used in 45 combinations using a complete block design. There were three densities of sowing on plots of 1.5 m^2 in area:

12.5 x 3.0 cm - sufficient for 100 kg grain/ha and approximately 400 plants/plot

20.0 x 3.0 cm - sufficient for 60 kg grain/ha and approximately 250 plants/plot

20.0 x 5.0 cm - sufficient for 40 kg grain/ha and approximately 150 plants/plot

and two shapes of linen isolators, one triangular isolator marked T, and the second rectangular and marked R. Control combinations, with unisolated plots, were marked K. Manual single grain sowing was performed. The previous crop was a legume mixture used as a green manure. Mineral fertiliser was applied at the rate of 40 - 50 kg N/ha, 90 kg P_2O_5/ha and 100 kg K_2O/ha.

Phenological observations were made for every field. The dates of sowing, emergence, shooting, heading, flowering, three maturation phases (milky, waxy and full) were noted and the periods of shooting - heading; heading - flowering; shooting - waxy maturity; emergence - full maturity were calculated. The following analysis of the components of the yield were performed:

- single spike grain yield (g)
- single plant grain yield (g)
- plot grain yield/1.5 m^2 (g)
- spike fertility (%) - percentage of filled spikelets compared to all
- 1 000 grain weight (g)
- total protein content by micro-Kjeldahl procedure on dry weight of grains (N x 5.83)

Measurements were made on ten plants from the middle of the plot and the results were then subjected to an analysis of variance. These tests were carried out separately because of different locations, three years in Skierniewice (1974 - 1976) and three years in Powsin experiments (1977 - 1979).

The rye materials grown at present in the Botanical Garden collection are also sown on plots of 1.5 m^2 at a density of 12.5 x 3.0 cm. Triangular isolators are used, protecting single rye varieties during flowering against cross-pollination by another variety. The density of 20.0 x 5.0 cm is used in regeneration of samples when only a small number of seeds is available.

RESULTS

Phenological phases

The data obtained proved that different densities of sowing and different methods of isolation have no influence on the length of single phenophases.

The measurements of phenological phases are presented in Table 1. In the Skierniewice experiment, the emergence phase in 1974/75 was late, and the vegetative period between emergence and full maturity was 25 days shorter than in 1973/74 and 32 days shorter than in 1975/76. Also, other phenophases contributing to the growth of green mass and grain formation were shortened.

TABLE 1

PHENOLOGICAL OBSERVATIONS OF RYE VARIETY DANKOWSKIE ZLOTE (1973/74 - 1975/76) AT SKIERNIEWICE (1976/77 - 1978/79) AT POWSIN

Season	Sowing	2 leaf emergence	Shooting	Heading	Flowering
1973/74	Sept. 29	Oct. 6	April 26	May 22	June 7
1974/75	Sept. 29	Oct 21	May 8	May 19	May 30
1975/76	Sept. 19	Oct. 2	April 29	May 23	June 8
1976/77	Oct. 4	Oct. 30	April 27	May 15	June 1
1977/78	Sept. 23	Oct. 13	May 1	May 24	June 5
1978/79	Sept. 22	Oct. 10	May 5	May 21	June 1

Season	Maturing phases			Phenological period (days)			
	Milky	Waxy	Full	Shooting - heading	Heading - flowering	Shooting - waxy maturing	Emergence - full maturing
1973/74	June 29	July 16	Aug. 12	26	16	81	300
1974/75	June 21	July 13	July 23	11	11	66	275
1975/76	July 3	July 24	Aug. 5	24	16	86	307
1976/77	June 26	July 18	July 28	18	17	82	272
1977/78	July 28	July 31	Aug. 6	23	12	91	298
1978/79	June 19	July 2	July 26	16	15	57	290

In Powsin, the late sowing in 1976/77 and then the late emergence caused a 26 day shortening of the whole vegetative period in comparison to 1977/78 and 18 days shorter than 1978/79. As a result, in 1974/75 and 1976/77, the plot yield of grains

TABLE 2

THE INFLUENCE OF DENSITY OF SOWING AND ISOLATION METHOD ON CHOSEN YIELD COMPONENTS OF RYE cv. Dańkowskie złote (Skierniewice 1974 - 1976)

Densities of sowing (cm)	Plot yield (g/1.5 m^2)				Single plant grain yield (g)			
	1974	1975	1976	mean	1974	1975	1976	mean
12.5 x 3 cm	980.20	252.33	907.67	713.40	3.84	1.03	5.39	3.42
20 x 3 cm	971.19	248.60	966.67	728.82	5.62	1.75	7.63	5.00
20 x 5 cm	867.27	207.00	870.00	648.09	8.50	1.27	9.19	6.32
Isolation type								
R	711.07	188.07	749.33	549.49	4.59	1.12	6.08	3.93
T	766.94	213.53	811.33	630.60	5.49	1.22	6.35	4.37
K	1240.66	306.33	1183.67	910.22	7.89	1.71	9.78	6.44
Mean	939.55	235.93	914.78	696.77	5.99	1.35	7.40	4.91
$LSD_{0.05}$*	60.52	48.87	120.04	47.22	0.54	0.24	1.06	0.39
$LSD_{0.01}$**	80.90	65.88	161.81	62.49	0.72	0.32	1.43	0.52

Densities of sowing (cm)	Single spike grain yield - mean (g)				Weight of 1 000 kernels (g)			
	1974	1975	1976	mean	1974	1975	1976	mean
12.5 x 3 cm	1.60	0.61	1.81	1.34	47.50	33.13	45.61	42.10
20 x 3 cm	2.21	0.70	2.22	1.71	48.71	31.57	47.13	42.47
20 x 5 cm	2.61	0.86	2.47	1.98	50.90	33.35	46.65	43.64
Isolation type								
R	1.70	0.59	1.94	1.41	48.63	33.11	47.05	42.93
T	1.98	0.69	2.07	1.58	48.32	31.73	44.81	41.62
K	2.72	0.88	2.49	2.03	50.23	33.22	47.53	43.66
Mean	2.14	0.72	2.16	2.17	49.07	32.68	46.46	42.74
$LSD_{0.05}$*	0.24	0.13	0.41	0.17	1.52	2.53	1.88	1.18
$LSD_{0.01}$**	0.32	0.18	0.55	0.23	2.03	3.41	2.54	1.56

* Least significant difference at 5% probability level

**Least significant difference at 1% probability level

TABLE 2 (cont.)

Densities of sowing (cm)	Spike fertility (%)				Protein content (% dry weight)			
	1974	1975	1976	mean	1974	1975	1976	mean
12.5 x 3 cm	70.26	67.69	73.82	70.59	11.66	11.39	13.15	12.88
20 x 3 cm	72.44	66.13	73.56	70.71	11.19	12.00	13.47	12.22
20 x 5 cm	72.91	65.85	73.97	70.91	11.51	12.35	13.79	12.55
Isolation type								
R	56.79	52.59	66.87	58.75	11.90	12.73	13.89	12.84
T	70.13	70.21	69.45	69.93	11.37	12.07	13.52	12.32
K	86.93	76.86	85.03	82.94	10.52	10.95	13.00	11.49
Mean	71.28	66.55	73.78	70.73	11.26	11.92	13.47	12.55
$LSD_{0.05}$*	4.34	6.36	4.39	3.01	0.26	0.58	0.63	0.30
$LSD_{0.01}$**	5.80	8.57	5.91	3.98	0.35	0.79	0.85	0.39

TABLE 3

THE INFLUENCE OF DIFFERENT SOWING DENSITY AND METHOD OF ISOLATION ON CHOSEN YIELD COMPONENTS OF RYE cv. Dańkowskie złote (Powsin 1977 - 1979)

Densities of sowing (cm)	Plot yield (g/1.5 m^2)				Single plant grain yield (g)			
	1977	1978	1979	mean	1977	1978	1979	mean
12.5 x 3 cm	336.64	818.75	554.73	570.04	1.88	3.36	2.24	2.49
20 x 3 cm	314.31	825.01	707.67	615.66	2.71	5.61	4.15	4.16
20 x 5 cm	311.13	837.94	614.74	587.94	3.69	8.90	5.30	5.96
Isolation type								
R	214.47	666.89	552.97	478.11	1.87	5.12	3.50	3.50
T	254.39	796.96	626.57	559.31	2.16	4.44	3.76	3.79
K	493.23	1017.86	697.60	736.23	4.25	7.31	4.43	5.33
Mean	320.70	827.24	625.71	591.22	2.76	5.96	3.90	4.21
$LSD_{0.05}$*	62.65	102.04	84.62	56.33	0.55	0.91	0.73	0.48
$LSD_{0.01}$**	84.46	137.55	114.07	81.93	0.74	1.23	0.98	0.70

*, ** Least significant difference at 5% and 1% probability level respectively

TABLE 3 (cont.)

Densities of sowing (cm)	Single spike grain yield - mean (g)				Weight of 1 000 kernels (g)			
	1977	1978	1979	mean	1977	1978	1979	mean
12.5 x 3 cm	0.63	1.05	0.67	0.78	32.76	43.73	35.42	37.30
20 x 3 cm	0.75	1.45	1.03	1.08	34.34	44.06	35.29	37.90
20 x 5 cm	0.83	1.97	1.14	1.31	34.80	47.38	35.26	39.15
Isolation type								
R	0.53	1.31	0.84	0.89	31.67	43.98	36.79	37.48
T	0.59	1.47	0.88	0.98	32.30	44.41	34.40	37.04
K	1.09	1.69	1.12	1.30	37.93	46.77	34.78	39.83
Mean	0.74	1.49	0.95	1.06	33.97	45.06	35.32	38.12
$LSD_{0.05}$*	0.13	0.30	0.21	0.18	2.43	1.59	2.98	1.90
$LSD_{0.01}$**	0.18	0.40	0.28	0.26	3.27	2.14	4.02	2.77

Densities of sowing (cm)	Spike fertility (%)				Protein content (% dry weight)			
	1977	1978	1979	mean	1977	1978	1979	mean
12.5 x 3 cm	59.56	74.08	84.05	72.70	11.20	9.92	11.28	10.80
20 x 3 cm	55.45	79.13	84.79	73.12	11.34	9.69	11.50	10.85
20 x 5 cm	52.57	76.64	83.29	70.93	11.58	10.06	11.67	11.10
Isolation type								
R	36.83	64.94	79.63	60.47	12.87	10.04	11.95	11.62
T	49.86	77.55	83.52	70.31	12.05	9.90	11.36	11.10
K	81.57	87.37	88.97	85.97	9.20	9.72	11.14	10.02
Mean	56.09	76.62	84.04	72.25	11.37	9.89	11.48	10.91
$LSD_{0.05}$*	7.51	6.78	3.30	4.73	0.89	0.55	0.44	0.43
$LSD_{0.01}$**	10.13	9.14	4.44	6.88	0.20	0.74	0.60	0.63

* Least significant difference at 5% probability level

**Least significant difference at 1% probability level

and, connected with it, the single plant and spike grain yield, 1 000 kernels weight and fertility of spikes were significantly lowered.

Spike fertility was strongly affected by the length of the isolation period. In 1978/79, the lack of rain, low humidity, and high temperature during flowering, made this phase very short and allowed us to take the isolators off the plants 5 days earlier, in comparison to 1977/78, and 10 days earlier than 1976/77. This minimised the influence of isolation on plants, but the effect was still significant in comparison to the unisolated control (K). Long term isolation of plants, because of rainy weather and prolonged flowering in 1976/77, caused a strong diminution of spike fertility, to 45.1% of the control in the case of R isolation and 61.1% of the control in T isolation.

The highly significant influence of different types of weather in different years on all measured features was proven through the three dimensional variance analysis of the 1974 - 1976 results.

Analysis of yield

The mean values of yield components for the 1974 - 1976 experiment are tabulated in Table 2, and for 1977 - 1979 in Table 3. The statistical analysis of those results are presented in Tables 4 and 5.

The differences in densities of sowing and in isolation significantly affected the single plot grain yield during the 1974 - 1976 experiment (Tables 2 and 4), but during 1977 - 1979, the densities did not influence the yield.

During 1974 - 1976, the highest single plot yield was produced at densities of 12.5 x 3 cm and 20 x 3 cm (the difference between them was not significant). The yield of grain from plots with a density of 20 x 5 cm was significantly lower.

During the 1977 - 1979 period, the lowest density plots at 20 x 5 cm compensated for the smaller number of plants per plot by an almost 100% increase in single spike grain yield, and the single plant grain yield (Table 3) in comparison to other densities of sowing. Because of this, there were no significant

TABLE 4

ANALYSIS OF THE VARIANCE: SKIERNIEWICE EXPERIMENT, 1974 - 1976

Source of variation	Degrees of freedom	Single spike grain yield - mean (g)		Single plant grain yield - mean (g)		Weight of 1 000 kernels (g)	
		Mean square (MS)	Variance ratio (F)	Mean square (MS)	Variance ratio (F)	Mean square (MS)	Variance ratio (F)
Years (Y)	2	30.73	180.81**	451.38	499.32**	3487.67	437.82**
Densities (D)	2	4.77	28.09**	94.85	104.92	29.27	3.67*
Isolations (I)	2	4.49	26.42**	81.01	89.62**	48.05	6.03**
(YxD) interaction	4	0.61	3.57*	13.84	15.31**	19.08	2.39
(YxI) interaction	4	0.53	3.14*	13.02	14.42*	4.85	0.61
(DxI) interaction	4	0.36	2.11	2.42	2.68	9.53	1.19
(YxDxI) interaction	8	0.42	2.48*	0.65	0.72	13.53	1.69
Error	104	0.17	-	0.90	-	7.96	-

Source of variation	Plot yield (g/1.5 m^2)		Total protein content (% d.w.)		Spike fertility (%)	
	Mean square (MS)	Variance ratio (F)	Mean square (MS)	Variance ratio (F)	Mean square (MS)	Variance ratio (F)
Years (Y)	7174445.25	560.57**	59.92	113.69**	606.44	11.67**
Densities (D)	82659.43	6.46**	4.99	9.80**	7.30	0.14
Isolations (I)	1611721.03	125.93**	20.98	41.18	6596.06	126.96**
(YxD) interaction	10795.79	0.84	0.38	0.75	19.43	0.37
(YxI) interaction	192479.16	15.04**	0.78	1.52	316.58	6.09**
(DxI) interaction	8346.11	0.65	0.10	0.20	2.68	0.05
(YxDxI) interaction	1955.67	0.15	0.59	1.17	27.84	0.53
Error	12798.46	-	0.51	-	51.95	-

* $F_{0.05}$ **$F_{0.01}$

plot yield differences between different densities of sowing of rye.

For both experimental locations, at Skierniewice (1974 - 1976) and at Powsin (1977 - 1979), plots with T isolators yielded significantly more grain than those with R isolators. The control plots significantly outyielded both types of isolated plots by about 50%.

The significant interaction between the years and isolations for both experiments means that there were differences in the relationships between the mean yield values for the 1975 harvest, in comparison to 1974 and 1976. The 1974 and 1976 isolations, T and R, did not differ significantly, in contrast to 1977 and 1979. Different densities of plants and differing type of isolation significantly influenced the single plant grain yield in both locations (Tables 2, 3, 4, 5). In 1974 - 1976, the largest grain yield from one plant was received from plots with the lowest density of plants (20 x 5 cm) and under T isolators. This was repeated again in 1977 - 1979, but there were no significant differences between both T and R shapes of isolators.

For the single plant yield, the interactions between years and densities were significant in both experiments.

Density x isolation and years x isolation interactions were significant only for 1974 - 1976 period. Years x density interaction was caused by the smaller influence on plant yield of different densities of sowing in 1975, being attributable to the difference between 12.5 x 3 cm and 20 x 5 cm densities. The same happened in 1977, compared with 1978 and 1979, causing the years x density interaction for 1977 - 1979 to be significant. The shape of the isolator affected the plant yield less in 1975 than in the other years.

The single spike grain yield was the largest at the lowest density of plants per plot (Tables 2 and 3). In both experiments, there was no significant effect of isolator type on grain yield per spike.

In 1974 - 1976 and 1977 - 1979, the years x density interaction was significant, and in 1974 - 1976, the years x isolation and years x density x isolation effects were also significant for spike grain yield. The years x density interaction for

TABLE 5

ANALYSES OF THE VARIANCE: POWSIN EXPERIMENT, 1977 - 1979

Source of variation	Degrees of freedom	Single spike grain yield - mean (g)		Single plant grain yield - mean (g)		Weight of 1 000 kernels (g)	
		Mean square (MS)	Variance ratio (F)	Mean square (MS)	Variance ratio (F)	Mean square (MS)	Variance ratio (F)
Years (Y)	2	1.35	49.49**	23.67	120.51**	329.25	107.96**
Density (D)	2	0.64	23.32**	27.09	137.89**	7.97	2.61
Isolations (I)	2	0.42	13.41**	8.72	44.40**	20.21	6.63*
(YxD) interaction	4	0.10	3.84*	2.91	14.81**	3.86	1.27
(YxI) interaction	4	0.02	0.90	0.61	3.10	13.56	4.44*
(DxI) interaction	4	0.03	1.18	0.70	3.58	0.69	0.26
Error	8	0.03		0.20		3.05	

Source of variation	Plot yield ($g/1.5\ m^2$)		Total protein content (% d.w.)		Spike fertility (%)	
	Mean square (MS)	Variance ratio (F)	Mean square (MS)	Variance ratio (F)	Mean square (MS)	Variance ratio (F)
Years (Y)	585344.73	218.76**	7.14	45.15**	1886.04	99.92**
Density (D)	4755.61	1.78	0.24	1.50	12.16	0.64
Isolations (I)	156782.35	58.59**	5.99	37.83**	1489.02	78.88**
(YxD) interaction	6961.79	2.60	0.04	0.28	23.67	1.25
(YxI) interaction	10764.41	4.02*	2.88	18.22**	272.83	14.45**
(DxI) interaction	3736.38	1.40	0.40	2.54	39.85	2.11
Error	2675.72		0.16		18.88	

* $F_{0.05}$

** $F_{0.01}$

1974 - 1976 was significant because of the different relations of mean values in 1975 compared with 1974 and 1976. In 1975, only the smallest density (20 x 5 cm) of plants differed significantly from others in spike yield.

For 1977 - 1979, significant interaction was caused by the different relations for the mean values in 1977, where T and R isolators did not differ significantly.

Similarly, for the interaction years x isolation in Skierniewice (1974 - 1976), the shape of the isolator did not significantly influence the spike grain yield in 1975.

For the years x density x isolations interaction in 1974 - 1976, there were no significant effects of densities of plants of the T and R isolators on spike yield in 1975.

The years 1974 and 1976 were similar, and in both, at the smallest density of sowing, the difference in spike yield for T and R isolators was not significant.

The spike fertility for both locations was significantly and positively affected by the T shaped isolation, as compared with the R type of isolation.

The most fertile spikes were, of course, produced by plots without any isolators (control) and after cross-pollination (control).

For spike fertility, the years x isolations interaction was significant (Tables 4 and 5), because of the significant differences between T and R isolators in 1974 and 1975, as well as in 1977 and 1978.

The 1 000 grain weight was affected in 1974 - 1976 to some extent by the isolation (Tables 2 and 4). The weight of 1 000 grains for T isolators was closer to K control and exceeded the R isolators, but significantly only in 1976.

The interaction for years x isolation was significant for kernel weight during the 1977 - 1979 period. The year 1979 was significantly different when the control plants did not differ from the isolated plants.

An increase of 1 000 grain weight was caused by the lowest density of sowing (20 x 5 cm) in 1974 and 1978, while plants from both the higher densities (12.5 x 3 and 20 x 3 cm) did not significantly differ in 1 000 grain weight at both locations at any time.

The protein percentage in rye grains did not depend significantly on the density of plants on plots in both locations. Therefore, the influence of isolation was significant. Isolated rye had, in comparison to the unisolated control, a significantly greater protein percentage in every year except 1978. The investigation of years x isolation effect for the 1977 - 1979 period was caused by significant differences in 1977 between the isolated and unisolated plants, as well as between both types of isolation (a smaller change in protein percentage of grains caused T isolators), while in 1978 and 1979 these differences were not significant.

For all of the measured features, a very strong variation was caused by year-to-year differences in climatic conditions (Table 4).

DISCUSSION

Plots of 1.5 m^2 in area are used in our collection for regeneration and multiplication of each rye taxon. Even at the lowest density of sowing (20 x 5 cm) the taxon is represented by 150 plants, which according to Geiger (1981), for conditions of free intercrossing provides an adequate population to ensure preservation of the gene pool of that heterogenetic taxon. At the highest density of sowing (12.5 x 3 cm) 400 plants represent each taxon.

Yields obtained during the experiment, from about 500 g/plot with rectangular isolators (mean), through about 600 g with triangular ones, to about 800 g for the control, were giving from about 13 000 to almost 20 000 grains per plot, which makes the multiplication rate range from 1 : 50 to 1 : 80.

The shortening of the vegetative period strongly decreased the mean spike, plant and plot grain yield, and often spike fertility, even when the shortening was due to late sowing and then the subsequent late emergence phase (in 1975 in Skierniewice and 1977 in Powsin). This confirms the importance of sowing the collection in time, and for a better comparison between the materials, sowing all the varieties in as short a time as possible. Because of the manual sowing of single grain this is very difficult to perform for such a large collection of material, but it should override the problems of organisation.

Another factor heavily influencing the yield components is the duration of plant isolation. It is clear that the shorter time inside an isolator with changed conditions (higher humidity and temperature, less light) results in a smaller difference in yield components compared with the unisolated control.

Therefore, it is very important to remove the isolator as quickly as possible, but only after flowering is completed. The length of the flowering period, apart from varietal differences, depends mainly on weather conditions. Rainy and cool weather prolongs this phase, as happened in 1977, greatly diminishing the grain yield. The yield of grain from each plot was not significantly affected by the two different densities of sowing, 12.5 x 3 cm and 20 x 3 cm. During one of the experiments (Skierniewice 1974 - 1976), the smallest density of sowing (20 x 5 cm) caused a significant diminution of grain yield from the single plot. Therefore, from the point of their yield, in both, a more dense sowing could be used. As was proved by Bruinsma (1966), both the above densities, sufficient to produce 100 and to 60 kg grains/ha, are optimal for winter rye. At lower seed rates, tillering was unable to compensate for the lower plant densities, while with seed rates over 100 kg/ha, the single spike grain yield decreased to a larger extent than the ear density increased.

In addition, at the lowest density of 150 grains/plot, in the case of grains received for sowing with a lowered germination capability, the number of plants/plot may be insufficient to represent the whole gene pool of the taxon.

To have as many plants as possible for better representation of the taxon, the highest density of sowing (12.5 x 3 cm) has been used.

Several times, the T shape isolator proved to allow better fertility of the spike than the R isolator, providing a higher grain yield from a single spike and single plant. This is probably due to the closer contact of the flowering spikes which are squeezed together under the top of the triangle isolator. Consequently, the tent shape T isolators are used in our collection at present.

The weight of grains, except for two years, was significantly affected by different plant densities, and, except in one year, by isolation. The differences in plant densities did not significantly influence the protein content (%) in grains, while the isolation of plants during the 1974 - 1976 experiment caused a significant increase in protein percentage in grains when compared with the control. In 1977 - 1979, grains from the control plants also had a lesser percentage of protein in all cases, but the differences were not significant. The triangle isolators always gave amounts of protein in grains which were closer to those in the control than in the rectangular shape of isolators. Therefore, for that reason also, the triangle isolators (T) seem to be better for future work.

We must bear in mind that prolonging the isolation of plants will affect the protein content of rye grains, but there is no other way, at the moment, to avoid cross-pollination in such a large collection.

Simultaneously, the heavy influence of weather conditions causes a high variability for all of the measured features from year to year (Tables 4 and 5) and in order to obtain certain results, several years of comparison with as extensive a collection of materials as possible are necessary.

Certainly, our collection grown in the same soil and weather conditions makes all our materials comparable between each other for the same year, all the more as the control variety, Dańkowskie Złote, is grown on every fifth plot.

CONCLUSIONS

1. Plots of 1.5 m^2 in area allow 150 to 400 plants to be grown which is enough to maintain the gene pool of heterogeneous taxons of *Secale*.
2. The date and duration of sowing of the rye collection is an important factor, heavily influencing the components of yield.
3. Isolation of rye during flowering influences the fertility of the spike, grain yield from the plant and plot, and the percentage of protein in grains, and therefore the isolation period should be as short as possible.
4. The density of sowing 12.5 x 3 cm was chosen as being best

for the rye collection based on plots of 1.5 m^2 because it ensures the best gene pool preservation of the defined taxon, even in cases of low germination.

5. Triangular (T) isolation created better conditions for pollination. The results were improved fertility of the spike, and a higher yield of grain from the spike, the plant and the plot. In addition, the grain protein content in these plants was closer to that of the control plants. Thus, the triangular (T) isolation has been chosen for further work in the collection.
6. Weather conditions were a great influence on yield components during the vegetative period, resulting in a need for several years of comparisons of numerous taxons from the rye collection in order to obtain certain results characterising these materials.

REFERENCES

Bruinsma J., 1966. Analysis of growth, development and yield in a spacing experiment with winter rye (*Secale cereale* L.). Neth. J. Agric. Sci., 14, 3, 198-214.

Brummund, M., 1974. Über Ergebnise von Proteinuntersuchungen an einer Roggensortiment Aktual. Probl. der Züchtungsfors und Zücht. bei Getreide, 127, 83-95.

Geiger, H.H., 1981. Breeding methods in diploid rye (*S. cereale* L.), EUCARPIA, Meeting of the cereal section on rye, Schwerin - GDR, June 22-27, 1981.

Gubanova, L.G. and Eroszenko, T.T., 1957. Mestnyje sorta ozimoj rzhi kak ishodnyj matjerial dlja kolekcji na uvielicenije soderzanija bielka. Vestnik sel´skochoz. Nauki, 9, 14-17.

Kotvics, G., 1970. Investigations on increasing the protein content of *Secale cereale* L., in: Protein growth by plant breeding. Ed. A. Balint, Akademiai Kiado, Budapest, 89-98.

Kubiczek, R. and Chojnacki, G., 1976. Characteristic of the nutritive value of the protein from rye caryopses. II. Evaluation of rye grain protein quality by Dye-Binding Capacity method. Acta Agrobotanica, 29, 2, 267-272.

Laube, W. and Quadt, K., 1959. Roggen (*Secale cereale* L.). Handbuch der Pflanzenzüchtung, 2, 6, 35-102.

Papp, Z., Szöke, S. and Ibranyi, K., 1956. Növenytermeles, 5, 3, 241-45 after Kotvics G., 1970.

Trzebska-Jeske, J. and Morkowska-Gluzińska, W., 1963. Zawartość azotu i niektórych aminokwasów niezbędnych w ziarnie żyta polskiego ze zbiorów 1959 r., Roczniki PZH, 14, 2, 153-164.

DISCUSSION

J.G. Hawkes *(UK)* Why do you think there are differences between the materials grown under rectangular and triangular isolators?

B.A. Molski *(Poland)* It is simply because there is better pollination in the triangular ones because the spikes are closer. Also there is a different microclimate.

J.G. Hawkes Have you tested this out?

B.A. Molski Yes. I don't have the figures with me but we have seen it in all the parameters.

W. Plarre *(Federal Republic of Germany)* Did you test the germability after you had isolated in different ways? I think this is very important. If you store this material, what is the longevity of the different isolations?

B.A. Molski There is no influence on germination in either the first or second years so it doesn't matter. The type of isolation does not influence the grain weight.

W. Plarre But you have different microclimates so therefore you could have a greater or lesser extent of contaminated seeds with fungus and so on. Are there any differences here because this is an influence on germability?

B.A. Molski Isolation is only for two or three weeks, during flowering time, not when the seeds are developing. There is no point in maintaining the isolation once pollination and flowering is over.

Lupin Workshop Lima, 12-21 April 1980
Recommendations of the working groups and some proposals for maintenance of different crops

W.PLARRE
Freie Universität Berlin, Germany

BREEDING

About twenty five people took part in this working group to discuss the different breeding problems that had arisen during the workshop. Lupin breeders and growers took the opportunity to consider some of these problems in greater detail and from different points of view. The principal intention was to determine how scientists could help the farmers, and to define the problems to be solved by scientific work.

The main conclusions are summarised in the following recommendations:

1. Genebanks

High priority is given for the exploration of centres of diversity in the Old World as well as in the New World. Developing aid programmes are urgently needed for collecting and storing germplasm. At the moment, lupins are registered in different national genebanks, for example, the Andean Crops Research Centre, CICA (Centro de Investigacion de Cultivos Andinos) located in the higher elevation of Peru, in Braunschweig (FRG) and elsewhere.

Lupins are becoming increasingly important, especially in South America. Therefore CICA must be supported by financing equipment and personnel on an international basis. CICA should be included in the network of the international genebanks. It has to take great responsibility in maintaining the basic collections of designated seed stocks for the whole Andean region. In order to determine the future role for CICA in co-operation with other genebanks and international centres, advice is required from the International Board for Plant Genetic Resources (IBPGR). Therefore, these recommendations should be sent to the secretariat and to national representatives of the genebanks, who are responsible for the co-ordination of the international programmes. The working group appreciates the activities of IBPGR. In their Annual Report for 1979, published under the title, 'Highlights of 1979', it is mentioned: "The way was paved for co-operative action .... particularly in South Asia, Meso-America and the Andean Zone".

Furthermore, in the Old World, especially in the eastern part of the Mediterranean region, new germplasm can be made available for breeding purposes. Research work in this field could be taken over by scientists of the Near East countries. Within EUCARPIA (European Association for Research on Plant Breeding), wild species and primitive forms are considered in section H, and the genebank committee should consider this. There is a good opportunity to discuss this matter during a business meeting at the EUCARPIA General Assembly to be held in Leningrad, 16 - 20 September 1980. As a Board member of EUCARPIA I have requested colleagues present to support this proposal in Leningrad. I will personally inform German representatives who will participate in the EUCARPIA General Assembly.

Concerning conservation of germplasm, it is recommended that new recorded accessions should be divided into two samples, to be stored and preserved in a second bank in such a way that the second sample is secure. There should be at least two international genebanks in which lupins are recorded, perhaps three (Andean Region, Germany, Near East, Southern Europe ?).

In order to discover the best solution for this problem, IBPGR should also work in collaboration with their national representatives.

2. Agricultural protected areas

The idea of protected areas to conserve not only the defined genotypes but also populations with continuing evolutionary processes, was introduced to the working group by Dr. Blanco (Peru). At present *L. mutabilis* is grown as populations in the higher elevations, where conservative farmers are working in a traditional way. However, agricultural progress moves on rapidly. If new cultivation methods and new cultivars are introduced, the old populations would disappear, leading to a gene erosion and the loss of valuable unknown genes. High priority should be given to prevent these processes; the only solution is to select some areas where the whole structure of traditional farming systems are still available and can be maintained. There will always be a more or less mild natural selection. The better adapted types would survive, and new mutants and recombinants with good fitness could be found. Such populations must be planted every year continuously.

This would involve a comprehensive agricultural programme under the leadership of an institute where breeders, geneticists, agronomists, pathologists, ecologists etc. are in collaboration. It is necessary to select a favourable ecological niche. A suitable place could be in the Cusco area.

As I know there are models which exist in Europe, it is recommended that the Agricultural Department of the Government of Peru should establish such an important programme. Details can be worked out by CICA. However, this is also an international task and therefore developing aid should be made available. There is a great need to protect crops as populations, not only lupins but also quinoa and others. We do not know which genes will be needed in the future.

TABLE 1

LUPINUS: SCALE OF CROSS-POLLINATION

L. polyphyllus (obligatory) > *L. mutabilis* > *L. albus* > *L. angustifolius* > *L. luteus*

TABLE 2

MAINTENANCE OF CROSS-POLLINATED CROPS

1. Released cv. or inbred lines by the breeder, special firms or organisations.
2. No more released cv.
   - 2.1 Old land races: in areas, in which they evolved (ecological niches). Mild selection pressure by environmental and anthropogenous factors = dynamic populations
   - 2.2 Older cv. no more in practice, in genebanks or institutes or in special farms, mass selection of repr. types = stabilised populations
   - 2.3 Inbred lines in genebanks

TABLE 3

MAINTENANCE OF GENETIC DIVERSITY

1. Self-pollinated crops
   - 1.1. By special breeding methods: composite crosses in connection with breeding goals, ♂ - sterility included (population improvement)
   - 1.2 Hybridisation, recurrent selection, different sub-populations
2. Cross-pollinated crops
   - 2.1 Hybridisation, recurrent selection, different sub-populations
   - 2.2 Single populations by top cross
   - 2.3 Synthetic populations by poly cross, cloning included
3. Maintenance of ecological niches (centres of diversity) for self- and cross- pollinated crops

DISCUSSION

J.T. Williams *(FAO/IBPGR)* I would just like to give a progress report because I am sorry to say that the recommendations of the working group in 1980 were somewhat out of date when they were made. The International Board for Plant Genetic Resources (IBPGR) has been collecting lupins in the Andes for at least three years, particularly in Peru, Bolivia and Ecuador, and will continue to work in Ecuador in association with IICA (Inter-American Institute for Agricultural Sciences). We have also been collecting in the Iberian Peninsula in association with the national programmes of Spain and Portugal, as part of the Med-iterranean programme, the latter in association with colleagues from the DDR. I regret to say that nothing has been done in the Near East, but a tremendous amount of collecting work has been done and all the material is available. We have not done any-thing in the Near East because we only put emphasis on crops of high global priority and as yet the lupin does not rank a high global priority. The IBPGR was dealing with 54 priority crops; this has now been revised and we have over 50 first priority crops so I doubt whether we will do very much in the Near East unless the countries themselves rank the crop to be of high priority.

In 1980 we also held a meeting on lupins in Spain; we finalised the descriptor list in association with European and South American colleagues and we are now proposing to designate base collections for lupins - two global collections, one in Gatersleben in DDR and one in a suitable genebank in the Andes when that will be built. In April of this year there was a meeting of the IBPGR, the countries of the Andean region in association with IICA and the Andean Seed Forum, where these matters were discussed. Since the Andean Seed Forum is an intergovernmental meeting, they are meeting again in October/November this year to take some decisions and it is highly likely that they will discuss an Andean genebank. So, there is quite a lot of activity taking place on lupins and all this material is freely available.

H. Lamberts *(Netherlands)* Thank you very much. I would add that in the Common Market group some of the countries have shown interest as well.

W. Plarre *(Federal Republic of Germany)* Dr. Williams, are the activities you have mentioned independent of the meeting in Lima?

J.T. Williams No. When the meeting was taking place in Lima we gave the background information to certain participants. IICA has already been involved in the work and I am a little surprised that they did not inform the meeting what was actually going on.

H. Lamberts Professor Plarre, are you thinking about seed production or green mass as well?

W. Plarre Primarily seed production, but lupins can also be used very well for green mass production, and perhaps bitter lupins of all kinds for green manure. I think there is also a market for this kind of production, especially in Europe.

Pollen dispersal in rye (*Secale cereale* L.): Research activities on isolation problems of cross-pollinated species of genebank material

Y.Z.KUTLU*
Aegean Regional Agricultural Research Institute, Menemen-Izmir, Turkey

INTRODUCTION

At the present time, thousands of seed samples of different plant species are stored in the genebanks of many countries. The objectives of these activities are:

1. To store the collected genebank material of different plant species for long-term usage, and
2. To distribute the material to the plant breeders.

In terms of regular distribution the quantities of the samples are reduced. Therefore, there is a necessity to multiply the material for future requirements. Since it is clear that there would be a decrease in the germination percentage, it is necessary to regenerate the samples. However, much more care should be taken to protect the genetic make-up of the material during multiplication and regeneration. In general, self-pollinated species of the crop plant can be regenerated and multiplied without isolation under open field conditions, whereas the cross-pollinated species are heterogeneous and heterozygous. Isolation is necessary during the flowering or pollination period. In wind pollinated species, control and regulation of pollination is especially difficult. Isolation by distance or isolation cabins is very expensive, in respect of the number of samples under multiplication. Rimpau (1882), and Heribert-Nilsson (1916), worked on this problem for their special breeding programmes. Later, Roemer (1932), Bateman (1947) and Kobabe (1965) also worked on these problems.

*English translation of excerpts from: Kutlu, Y.Z., Die Ausbreitung von Pollen beim Roggen (*Secale cereale* L.), from Z. Pflanzenzüchtung, 78, 253-264 (1977). Paul Parey, Berlin.

In this study, rye *(Secale cereale* L.*)* was chosen as a model plant to represent the group of typically wind pollinated species. Three different methods were used to measure the dispersal of pollen grains and to determine the rate of cross-pollination:

1. Measurement of pollen dispersal by means of pollen tracers and microscopic analysis.
2. Measurement of cross-pollination by means of self-sterile clones of perennial rye as indicator plants using genetic analysis.
3. Measurement of pollen dispersal by means of the Indicator Activating Method (IAM).

1. <u>Microscopic analysis of pollen dispersal</u>

Kustro rye in a 0.6 ha field, was used as the pollen source. All three methods consisted of a central source of donor pollen while the measuring devices (pollen tracers and indicator plants) were placed at various intervals in eight rows of plants radiating from the central pollen source. This experimental design was found to be suitable since the dispersal of rye pollen is particularly explosive and closely related to the direction and strength of the prevailing wind. The pollen tracers were changed every day at the same time during the pollination. At the end of pollination the rye pollen was identified and counted with the assistance of a light microscope by 65 x enlargement.

The pollen tracers were wooden cylinders, 2.5 cm in diameter and 6 cm in length. Polyethylene sheets 4 x 8 cm were attached to the cylinders to catch pollen grains by static electricity.

2. <u>Genetic analysis of cross-pollination</u>

In this work a self sterile rye clone was used to determine cross-pollination rate. The self-fertility of this clone which was received from Professor Hondelmann in October 1973, was only 0.03% (Reimann-Philipp et al., 1968, Hondelmann and Sneyd, 1973). One hundred and sixty plants were derived by preparation of tillers followed by planting in plastic pots.

The cloned plants were placed in the open field during the winter season to vernalise (Bommer, 1967; Sneyd, 1975). During pollination, the plastic pots were used as pollen tracers in 8 directions up to a distance of 1 000 m. After ripening the plants were threshed and the grains counted. The numbers of ears produced per pot ranged from 10 - 27.

3. Indicator Activating Method (IAM)

The element Dysrosium 63, which is rarely found in nature, was used to mark the rye pollen grains. This method of Glubrecht (1961) is different from the normal activating method of Fomin et al. (1957) and Sdrilko et al. (1958). In this study there was no release of radioactivity during the field experiments, Dy-63 remaining stable in the pollen grains until neutron activation in an atomic reactor.

The summer rye variety Somro was planted in 40 Kick-Brauchmann pots, each of which was supplied with 300 mg of Dy-63 in a nutrition solution. These pots were placed in the field before pollination. These 40 pots were considered equivalent to the standard multiplication plots in the German genebank at Braunschweig-Völkenrode with an area of 6.4 m^2. Pollen tracers were placed at various intervals in eight rows up to 50 m distant from the central pollen source.

After pollination and neutron activation, the radioactivity of 1 000 pollen grains was measured, and that of each pollen grain determined. All the pollen tracers were activated in an atomic reactor at the Physical Technical Bundesanstalt (PTB) and after activation radioactivity was measured at the Isotope Laboratory of FAL (Forschungsanstalt für Landwirtschaft). The number of pollen grains was determined after the measurement, under controlled conditions without free pollination.

Statistical analysis

1. Pollen dispersal at various distances was analysed by regression analysis.
2. Pollen dispersal in eight wind directions was evaluated by an analysis of variance.

RESULTS

1. Microscopic analysis

Average values indicated that pollen was dispersed over a greater distance in 1974 than in 1975. Also the average number of pollen grains caught by the tracer was more in 1974 (4 514) than in 1975 (2 807). This was attributed to the wind speed during pollen dispersal (in 1974: 4.29 m/sec and in 1975: 3.37 m/sec). Significant differences were found in 1974, except between the directions W-NW, N-S and E-SE. In 1975, prevailing NE winds produced a strong directional SW effect. In addition to the wind, temperature, humidity and rainfall are also important in pollen dispersal.

The correlation coefficients of these factors for both years were as follows:

Wind speed :	0.43	($P > 0.1$)
Rainfall	- 0.25	($P > 0.1$)
Humidity	0.14	
Temperature	0.03	

2. Genetic research

The number of grains per ear was counted in both seasons (1974 and 1975) and values were found to be significantly different at various distances. Differences between individual wind directions in both years were also significant, the percentage of cross-pollination being highest in the W direction, because of the prevailing E wind.

3. Indicator Activating Method

One thousand pollen grains were taken from the first and second experiments to determine the amounts of radioactivity in each of the pollen tracers. The radioactivity of a single pollen grain was then calculated. The results were as follows:

9.25×10^{-8} μg Dy/pollen grain from the first experiment

13.25×10^{-8} μg Dy/pollen grain from the second experiment

Ninety seven percent of the total pollen grains were found at 20 m distance in the first experiment. There were no pollen grains at 50 m distance. Eighty four percent of the total pollen grains were found at 3 m distance in the second experiment.

It was not necessary to make any statistical analysis for the wind directions, because of the prevailing W wind with both experiments.

REFERENCES

Bateman, A.J., 1947. Contamination of seed crops. III. Relation with isolation distances. Heredity 1, 303-336.

Bommer, D., 1967. Photoperiodic response and vernalization in grasses. Rep. Meeting Eucarpia Foddercrops Sect., Köln-Vogelsang, Germany, Sept. 20th and 21st, pp. 71-81.

Fomin, A.Je. and Asstachowa, N.K., 1957. Zur Methodik der Pollenmarkierung bei Pflanzen (russ.) Selekcija i semenovodstvo 22 (3), 72-73.

Glubrecht, H., 1961. Die Indikator-Aktivierungsmethode. Atompraxis 7, 467-470.

Heribert-Nilsson, N., 1916. Populationsanalysen und Erblichkeitsversuche über die Selbststerilität, Selbstfertilität und Sterilität bei dem Roggen, Z. Pflanzenzüchtg. 4, 1-44.

Hondelmann, W. and Sneyd, J., 1973. Perennierender Kultur-Roggen. I. Gegenwärtiger Stand der Züchtung. Z. Pflanzenzüchtg. 69, 89-101.

Kobabe, G., 1965. Einfache Hilfsmittel für die Bestäubungsregulierung bei der Züchtung von Fremdbefruchtern. Züchter 35, 299-307.

Reimann-Philipp, R. and Rohde, H., 1968. Die zytologische Identifizierung der genetisch unterschiedlichen Gruppen von Artbastarden in den späteren Generationen der Kreuzung *Secale cereale* x *Secale montanum* in ihrer Bedeutung für die Züchtung eines perennierenden Kulturroggens. Z. Pflanzenzüchtg. 60, 212-218.

Rimpau, W., 1882. Das Blühen des Getreides. Landw. Jb. 11, 895-911.

Roemer, Th., 1932. Über die Reichweite des Pollens beim Roggen. Z. Pflanzenzuchtg. 17, 14-21.

Sdrilko, A.F. and Chawshinskaja, O.J., 1958. Zur Frage der Wege des Eindringens von 35S und 32P aus markierten Pollen in die Fruchtknoten (russ.). Bull. Ukr. naucno-issledov. Inst. Rast. Sel. i Genet. 3, 27-30.

Sneyd, J., 1975. Ertragsphysiologische Untersuchungen zur Nutzung von perennierenden Roggenformen im Pflanzenbau. Diss. Giessen, 175 S.

DISCUSSION

E. Porceddu *(Italy)* I think this paper was very helpful to our meeting here because almost all the speakers referred to random mating and very few considered the fact that probably the pollen does not spread in all directions, at random. Such things as wind direction and the possible preference of some insects for certain coloured flowers, etc. are relevant factors. This paper clarifies a number of points that have been touched on during the meeting.

Final session

Chairman: E.PORCEDDU

Recommendations of the Working Groups

The meeting was divided into Working Groups to consider and formulate Recommendations. Draft recommendations were circulated to all delegates and were subsequently discussed and amended as necessary at the Final Plenary Session of the Symposium.

The Final Recommendations were as follows:

GENERAL

National parks in EC member states should include plans for conserving gene pools and old ecosystems.

EUCARPIA sections (cereals, fodder crops etc.) and EUCARPIA groups (*Medicago sativa*, etc.) should participate in all the aspects of germplasm conservation, preparing specific practical procedures.

Old cultivated populations and garden varieties are often removed from official registered lists (EC and state lists). This is an important cause of genetic loss and the material should be automatically offered to genebanks for conservation.

It is accepted that an accession is accompanied by a taxonomic name which helps to define it in the best possible way. A check on whether the name is correct needs to be indicated in data banks. For botanical taxa, this check has to be made by comparing material with at least the original description. A descriptor and its state should be added.

Similarly, with cultivars, a check has to be carried out. A descriptor and its state can be added using one or more digits to include checks against living standard, conserved standard, original description and/or experience of experts.

The checks should be repeated after each regeneration of base collections because contamination, drift and shift of frequencies of character can occur, thereby changing the accession at that stage.

CO-OPERATIVE PRESERVATION PROGRAMME

For genetic preservation of cross-fertilised species, various methods like *in situ* preservation, collection centres, the *in vitro* method and seed storage are available. However, as none of these is the perfect method, a co-ordinated research should be organised in order to determine the range of variability and detect the main defects in each method. This work should be done on model crops such as lucerne, maize, *Lolium*, rye, *Brassica* etc. because the largest genetic diversity exists in Europe and it is a source of material for all breeders of the world.

The following steps are recommended to achieve this objective.

i. Determine agricultural areas with specific agro-ecosystems which will be designed as cultivar reservation parks. (Some farmers must be supported to maintain these cultivars or populations).

ii. Multiply the populations in an artificial environment comparing synthetics, chain-cross, grouping etc.

iii. Promote the most accurate methods to analyse and identify the genetic structure, allelic frequencies, phenotypes composition etc. of the populations (biometry, discriminant analysis, electrophoresis, immunology).

iv. Encourage the establishment of a tissue culture laboratory devoted to research on meristem and gametic storage and regeneration.

v. Conduct comparative trials between different methods after every five years or after experimental seed deterioration to determine the specificity of genetic drift for each method and the possibility of correcting one method by another.

A small group comprising experts in plant population genetics and plant breeding and related fields should supervise the agenda and co-ordinate the activities of this project.

TRAINING

It is proposed to organise training courses in addition to those already offered because the management of genetic resources requires broad basic knowledge in the field of genetics and plant population dynamics and also to bring about closer understanding between plant breeders and experts in genetic conservation.

These training courses could be devoted mainly to genetic basis, agronomy, methodology of genetic conservation and breeding needs of genetic resources for breeding programmes.

As many French-speaking North African and tropical African countries would like to be involved, these training courses could be conducted in French with participation of specialists from the EC.

FORAGE CROPS

We affirm that the primary objectives of genebanks are:-

i. to extend the genetic resources available to breeders through the collection and storage of diverse plant material

ii. to collect and preserve genetic material threatened with erosion through changes in agricultural practice and land use.

The aim is to preserve, as far as possible, the genetic *status quo* of each accession (collection) by the best available means so that the identity and integrity of each collection (ecotype) is preserved as a genetic resource for future exploitation by breeders and researchers.

Presently, the most practical method would appear to be the random pollination of representative individuals of each collected population under controlled conditions (using principles formulated for the multiplication of synthetic varieties). The salient considerations are:-

a. Isolation - to avoid contamination from foreign pollen.

Isolation distances vary with the species and the environment. For some wind-pollinated species, a distance of 200 m for small stands is recommended. Relatively safe distances of the same order are suggested for insect-pollinated species. Clearly, more research is required in this area.

b. Minimum number of plants to avoid genetic drift - The critical number depends on the genetic structure of the population and precise aims. But in lucerne, the population structure is such that a minimum number of 200 is proposed. For ecotypes of outbreeding grasses and clovers, an absolute minimum of 25 plants should be used to avoid drastic genetic drift, but there are strong arguments for increasing this number to 50 or more if practical.

c. <u>Avoiding shift through natural selection</u> - This is the most hazardous, under-researched and controversial aspect. Regeneration in the area of origin has many attractions. Failing this, selective elimination of the representative genotypes is avoided by growing the plants under controlled conditions, while maximum seed production from each genotype is ensured by growing the plants at low densities.

Seed is collected from each genotype individually and is then mixed in equal proportions so as to obtain equal representation of the genotypes in the basic seed which is to be used for storage and regeneration. Seed from single plants could be handled separately.

<u>Storage</u> - The aim is to produce seeds of high quality of a sufficient quantity for evaluation, seed exchange and storage needs. Optimum conditions for seed storage have been discussed elsewhere.

<u>Research requirements</u> - Most of the areas of consideration require further research. We identify some of those which are particularly imperative.

1. The differential effects of ageing in seed from different genotypes with regard to (i) viability (physiological effects), (ii) genetic deterioration in surviving plants.
2. Competitive interactions between genotypes, especially a comparison of genetic change with increasing densities during reproduction.
3. Extent of drift and shift under different systems of regeneration.
4. More effective and cheaper isolation cages and houses.
5. Finally, it is felt that for perennial forage grasses and some legumes which have evolved and are maintained by selection pressures brought about by livestock systems, due consideration should be given to the extent that conservation may be obtained *in situ*, in addition to or complementary with long-term seed (or vegetative) storage.

Training - The group recognises the need for training of practitioners in these disciplines. This may be particularly imperative in certain regions of the world.

HORTICULTURAL PLANTS

The conference underlined the primary importance of conserving the genetic integrity of the population sample from generation to generation with minimum loss or change of gene frequencies.

When it becomes necessary to regenerate the stored seeds, it is essential to prevent genetic drift, inbreeding depression and the loss of heterozygosity. The method of accomplishing this is to ensure that not less than 25 and preferably 50 to 100 plants should be used.* Under exceptional circumstances, fewer plants may be used. Information should be available on the number of plants used and the methods of regeneration. Information should also be available on the number of regeneration cycles through which the material has passed since it was collected.

It is clear that extreme care must be taken by means of established techniques to prevent pollen contamination.

Suggested storage figures are as given on page 29 of the Crop Genetic Resources Field Collection Manual.

Users who request samples from a genebank should be responsible for multiplying these for their own purposes.

Users should indicate the purpose for which the seeds are needed and the numbers required.

In view of the interest expressed in the use of insects for controlled cross-pollination, a small workshop on this topic should be arranged in the near future.

FIELD CROPS

The discussion concentrated on the practical aspects as presented by De Pace at this meeting.

The group suggested that as far as *Secale cereale* is concerned, prevention of pollen flow should be defined as isolating at least 100 plants.

*Population size and effective population size are not necessarily the same.

The group felt that more information should be gathered on the biology of reproduction and a deeper knowledge on population structure should be obtained. Practical information is already available. (See Table 1 for examples).

There is a need for a special conference, workshop or other training possibilities on population genetics of cross-pollinated species especially in relation to collection and seed regeneration.

For the maintenance of land races and ancient crops, consideration should be given to the *in situ* conservation in protected areas with special ecological niches.

TABLE 1

PRACTICAL ASPECTS

	Zea mays	*Helianthus annuus*	*Vicia faba*	*Secale cereale*
Population size	20-100 plants well-spaced.	20-100 plants well-spaced.	20-80 plants well-spaced.	100-200 plants well-spaced.
Pollen flow prevention	Use paper bags to cover randomly chosen spikes.	Use bags to isolate randomly chosen inflorescences.	100 m distance between plots or less if artificial or natural (plant) barriers are present.	100 m distance between plots or isolation cages or forming at least ten groups of ten plants and bagging together spikes from plants of the same group.
Mating system	Artificial intercrossing 20-80 female plants with pollen pool from all the others	Artificial intercrossing 20-80 female plants with pollen pool from all the others.	Allow insect pollination.	Natural intercrossing among neighbouring plants.
Harvesting	Same number of seeds; about 100 from each spike.	Same number of seeds; about 100 from each inflorescence.	Same number of seeds; about 10 from each plant half from apical, and half from basal nodes.	Same number of seeds; about 10 from each plant.

Closing

E. Porceddu *(Italy)* The meeting has now come to a close. We had four invited papers and fifteen voluntary contributions, six of which were from non-EC countries. This reaffirms the fact that science has no frontiers and that, among us, there is a deeper friendship than politicians perceive.

The participation of non-EC scientists was by courtesy of EUCARPIA and I wish to thank Dr. Lamberts, the President and Professor Hawkes, the Chairman of the session 'Wild Species and Primitive Forms' for providing a large umbrella under which scientists have a common interest.

The meeting also provided a good opportunity for scientists of different disciplines to present their ideas and submit them for discussion. I am sure that these ideas, discussions and resolutions will contribute to the enrichment of our knowledge and thinking.

This will also immensely benefit the different groups of scientists here, geneticists, breeders, botanists and those involved more directly in the day to day management of genebanks.

I think this is the first meeting of its kind in Europe, and probably one of the first in the world, on conservation of genetic resources of cross-pollinated species, a group of plants which represents a large part of the agricultural output and provides staple food for people of different regions. We are all convinced that this natural resource has to be preserved in the best way for future generations.

Every day we hear about the problems and crises regarding oil, energy etc. but very little is heard about agricultural aspects and genetic resources which, as suggested by Jacob and his co-workers in their report to the President of the French Republic, will represent an indispensable resource for the human race during the next millenium.

It is with this idea in mind and our responsibility to future generations that I thank all of you for your participation and for having provided stimulating ideas. I hope we shall have more opportunities to meet in the future to improve the procedures of preservation of this invaluable heritage that past generations have left to us, and to strengthen our friendship.

List of participants

Dr. S. Andersen
Dept. Crop Husbandry and
Plant Breeding
Thorvaldsensvej 40
Copenhagen V
DENMARK

Dr. D. Astley
National Vegetable Research Station
Wellesbourne
Warwick CV35 9EF
UNITED KINGDOM

Dr. Z. Athanassios
Cereal Institute
Thessaloniki
GREECE

Dr. G. de Bakker
Executive Secretary
FAO/UNDP Programme for the
Conservation of Genetic Resources
(ECP/GR)
Palais des Nations
1201 Geneva 10
SWITZERLAND

Dr. A. Boggio
Institut de Recherche Vilmorin
La Menitré
49250 Beaufort-en-Vallée
FRANCE

Dr. H.H. van der Borg
Ministry of Agriculture
and Fisheries
Directorate of Agricultural
Research
Mansholtlaan 4
6708 PA Wageningen
NETHERLANDS

Dr. W.A. Brandenburg
Dept. Taxonomy of Cultivated Plants
Agricultural University
Haagsteeg 3
6708 PM Wageningen
NETHERLANDS

Dr. E.L. Breese
Welsh Plant Breeding Station
Plas Gogerddan
Aberystwyth
Dyfed SY23 3EB
UNITED KINGDOM

Dr. V. Connolly
Agricultural Institute
Oak Park Research Centre
Carlow
IRELAND

Miss. H.A. Cortessi
Hellenic Sugar Industry
PO Box 108
Thessaloniki
GREECE

Prof. Y. Demarly
Amélioration des Plantes
Bâtiment 360
Université Paris-Sud
91405 Orsay
FRANCE

Dr. P. Guy
INRA
Station d'Amélioration
des Plantes Fourragères
86600 Lusignan
FRANCE

Prof. J.G. Hawkes
Department of Plant Biology
University of Birmingham
PO Box 363
Birmingham B15 2TT
UNITED KINGDOM

Dr. L. Holly
Research Centre for Agrobotany
NIAVT
H-2766 Tapioszele
HUNGARY

Dr. J. Jakubiec
Institute of Genetics
and Plant Breeding
University of Agriculture
ul. Nowoursynowska 166
02-766 Warsaw
POLAND

Mr. G. Jenkins
Agricultural Research Council
160 Great Portland Street
London W1N 6DT
UNITED KINGDOM

Dr. K. Kostov
Institute for Plant Introduction
and Genetic Resources
Sadoro
BULGARIA

Dr. Aune Kurki
Nordic Gene Bank
Box 1543
22101 Lund
SWEDEN

Dr. Y.Z. Kutlu
Aegean Regional Agricultural
Research Institute
PO Box 9
Menemen-Izmir
TURKEY

Dr. R. Laeremans
Katholieke Universiteit Leuven
De Croylaan 42
B-3030 Leuven
BELGIUM

Dr. H. Lamberts
Director
Foundation for Agricultural
Plant Breeding
SVP
PO Box 117
6700 AC Wageningen
NETHERLANDS

Dr. C. Lehmann
Central Institute of Genetics and
Cultivated Plant Research
Academy of Sciences of the GDR
DDR-4325 Gatersleben
EAST GERMANY

Dr. B.A. Molski
Director
Botanical Garden
Polish Academy of Sciences
Powsin
02-973 Warsaw 34, skr. 84
POLAND

Dr. L.M. Monti
Facoltà Agraria
Università di Napoli
80055 Portici
ITALY

Dr. Katarzyna Niemirowicz-Szczytt
Institute of Genetics
and Plant Breeding
University of Agriculture
ul. Nowoursynowska 166
02-766 Warsaw
POLAND

Dr. C. de Pace
Plant Breeding Institute
University of Bari
Via Amendola 165 A
70126 Bari
ITALY

Dr. C.P. Pedersen
Daehnfeldt Ltd.
PO Box 185
DK-5100 Odense C
DENMARK

Dr. P. Perrino
Istituto del Germoplasma
Via Amendola 165 A
70126 Bari
ITALY

Dr. H.L. Petersen
Department of Crop Husbandry and Plant Breeding
University of Agriculture
Thorvaldsensvej 40
DK-1871 Copenhagen V
DENMARK

Dr. W. Plarre
Institut Angewandte Genetik
Albrecht-Thaer-Weg 6
D-1000 Berlin 33
FEDERAL REPUBLIC OF GERMANY

Dr. E. Porceddu
Istituto del Germoplasma
Via Amendola 165 A
70126 Bari
ITALY

Dr. M.H. Poulsen
Department of Crop Husbandry and Plant Breeding
University of Agriculture
Thorvaldsensvej 40
DK-1871 Copenhagen V
DENMARK

Dr. Luciana Quagliotti
Institute of Plant Breeding and Seed Production
Via Giuria 15
10126 Turin
ITALY

Dr. Henny Roelofsen
Institute for Horticultural Plant Breeding
Mansholtlaan 15
Wageningen
NETHERLANDS

Dr. R. Salvo
Department of Crop Husbandry and Plant Breeding
University of Agriculture
Thorvaldsensvej 40
DK-1871 Copenhagen V
DENMARK

Mr. R.D. Smith
Royal Botanic Gardens
Wakehurst Place
Ardingley
Haywards Heath
W. Sussex RH17 6TN
UNITED KINGDOM

Dr. B.H. Somaroo
Program Leader Forage Improvement
ICARDA
PO Box 5466
Aleppo
SYRIA

Dr. P. Spagnoletti-Zeuli
Istituto del Germoplasma
Via Amendola 165 A
70126 Bari
ITALY

Dr. J. Tammisola
Dept. of Plant Breeding
University of Helsinki
00710 Helsinki 71
FINLAND

Mr. B.F. Tyler
Welsh Plant Breeding Station
Plas Gogerddan
Aberystwyth
Dyfed SY23 3EB
UNITED KINGDOM

Dr. F. Veronesi
Centro di Studio per il Miglioramente Genetico delle Plante Foraggere del CNR
Istituto di Allevamente Vegetale
Borgo XX Giugno
06100 Perugia
ITALY

Dr. J.T. Williams
Executive Secretary
International Board for Plant Genetic Resources
Food and Agriculture Organisation of the United Nations
Via delle Terme di Caracalla
00100 Rome
ITALY

Dr. F. Yndgaard
Deputy Director
Nordic Gene Bank
PO Box 1543
22101 Lund
SWEDEN

EC

Dr. C. Caruso

Dr. J. Dehandtschutter